P9-DVC-168

CHARGING AHEAD

Other books by the author

In the Rings of Saturn
Fast Lane on a Dirt Road
A Thousand Voices

CHARGING AHEAD

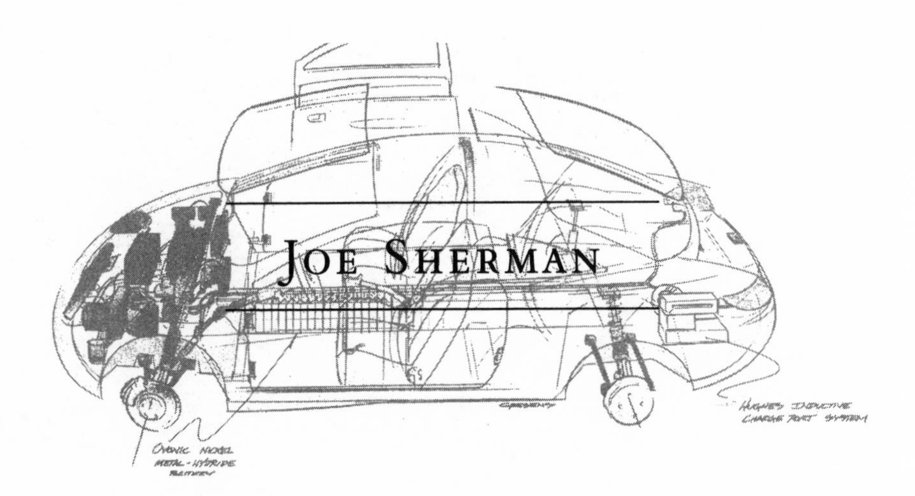

JOE SHERMAN

New York Oxford
Oxford University Press
1998

Riverside Community College
Library
NOV '99
4800 Magnolia Avenue
Riverside, CA 92506

TL 220 .S54 1998

Sherman, Joe, 1945-

Charging ahead

Oxford University Press

Oxford New York
Athens Auckland Bangkok Bogotá
Buenos Aires Calcutta Cape Town Chennai Dar es Salaam
Delhi Florence Hong Kong Istanbul Karachi
Kuala Lumpur Madrid Melbourne
Mexico City Mumbai Nairobi Paris São Paulo Singapore
Taipei Tokyo Toronto Warsaw

and associated companies in
Berlin Ibadan

Copyright © 1998 by Joe Sherman

Published by Oxford University Press, Inc.
198 Madison Avenue, New York, New York 10016

Oxford is a registered trademark of Oxford University Press, Inc.

All rights reserved. No part of this publication may be reproduced,
stored in a retrieval system, or transmitted, in any form or by any means,
electronic, mechanical, photocopying, recording, or otherwise,
without the prior permission of Oxford University Press.

Library of Congress Cataloging-in-Publication Data
Sherman, Joe, 1945–
Charging ahead / Joe Sherman.
p. cm.
Includes bibliographical references and index.
ISBN 0-19-509479-4
1. Automobiles, Electric. I. Title.
TL220.S54 1998
629.22'93—dc21 98-23216

1 3 5 7 9 8 6 4 2

Printed in the United States of America
on recycled acid-free paper

this book is dedicated to
Nancy Hazard
for her tireless work on behalf
of green cars and a
sustainable future

Cleaning, in planetary terms, and unlike other sectors, held wonderful promise. Cleaning was obviously going to be huge.

—Martin Amis, *The Information*

Contents

Preface xi

1. The Freewheeling James Worden Wins Another Race 2
2. Then and Now: Rebirth of a Dormant Industry 10
3. James Testifies on the Promise of Clean Cars 18
4. An Idea Whose Time Has Come? 22
5. James's Second Car: A Goofy-Looking Aluminum Thing 26
6. Racing Solar-Electric Cars at MIT 30
7. Air: This Marvelously Tattered Canopy 34
8. Australia, 1987: The First World Solar Challenge 48
9. 1989: The Founding of Solectria Corporation 52
10. Forces Gathering Behind Cleaner Cars and Air 58
11. 1989–1992: Solectria's First Years in Business 66
12. A Regulatory Minefield: The Ozone Transport Tale 76
13. "No Zero-Emission Vehicle Mandate Is a Stake In the Heart" 94
14. The Sunrise Project and Its Partners 98
15. Anaheim, 1994: Twelfth International Electric Vehicle Symposium 122
16. Sunrise Almost Stalls, Then Rolls, Finally Races 134
17. May 1995: The Seventh NESEA American Tour de Sol 154
18. Crippling the Clean-Car Mandate 164
19. Getting Sunrise into the Mainstream 174

Epilogue 187

Notes 193

Index 215

Preface

James Worden, the young CEO at the center of this book, classifies himself, when he classifies himself at all, as an efficiency guy, one with a love for cars and for efficient technologies. He's also, in his own words, "a sticker." That is, a guy who stays with a job, with a dream. Worden's dream is to build and sell the cleanest, most efficient car technologies in America, in the world.

This book describes some of his efforts to achieve that goal between the mid-1980s, when he was making electric cars in his parents' backyard, to the late 1990s, when his company, Solectria Corporation, was seeking a manufacturing partner to put Sunrise, an all-composite, lightweight, highly efficient four-passenger electric sedan for which Solectria had developed many new techologies, into production.

It's not a personal look at Worden. Readers won't come away knowing his sex life or how much money he makes or what he dreams about at night. It's about the cars he built, their technological evolution, the start-up and growth of his youth-driven, dream-seeded company. It's about that classic American story: getting a company off the ground without having it collapse on top of you. Finally, it's about the context, the environmental and regulatory atmospheres, that have made Worden's automotive journey not only possible but important.

The world's air is in trouble. Think about it. Every breath you take carries that message, especially if you live in a large metropolitan area, as most Americans do. Global warming is a collective atmospheric anxiety, beyond individuals, beyond countries, beyond the world's willingness to confront it, although that is gradually changing as scientific evidence accumulates validating its reality.

The two biggest industries in the world, autos and oil, don't much like people worrying about air pollution and global warming. That's easy to understand. The two industries cause much of both. Of course, people everywhere want their cars and trucks, and burn oil in houses, factories, offices, and elsewhere, so the industries rationalize emissions and air problems with the logic that they're only giving people what they want. Which is right, to a certain degree. But auto and oil marketing promote cars and oil and their public relations and legal arms fight alternatives, and those who champion alternatives, with deep pockets and blindered vision. All too often, it seems to me, the industries, through dint and cash and political fear, prevail. Bottom line? Things stay the same: air gets dirtier, global warming becomes more problematic, alternatives get sidetracked.

In this story I attempt to show how the development of electric cars in America, between 1988 and 1996, almost broke that pattern. And how James Worden and his company rode that optimistic era, when it looked as though not only electric cars would appear on the roads in greater numbers but would be made by companies other than GM, Ford, and Chrysler, by smaller, regional companies allied with technology firms. And how ultimately the Big Boys, as the auto industry is called, got their act together to kill regulations that were pushing cleaner, more efficient cars into the marketplace soon and to eliminate many of the small companies whose survival hinged on the regulations.

Solectria avoided being eliminated from the electric vehicle industry. Compared to many of the other small startups in the new industry, Worden's company was more agile, smarter. It cultivated the right allies, not the least of which, in an ironic twist, was the Pentagon. After the Persian Gulf War, the Pentagon decided it wanted to see electric vehicle technologies in battlefield vehicles. So defense money poured into EV developments, and Solectria ended up being a recipient of the military's largess.

To give the automakers credit, they gradually did concede to the environmental and technological logic behind EVs, but only after they became confident they were killing the regulations forcing the new technologies forward. In the end the automakers gained control of the rate of change, which is what they wanted. Solectria survived but, at this writing, seeks a partnership with one of the automakers in order to get its advanced technologies into the mainstream.

Writing this book with Solectria at its center would have been impossible without the cooperation of James Worden, Anita Rajan Worden, and many Solectria employees, among them Ed Trembly, Arvind Rajan, Karl Thidemann, Wayne Kirk, Mark Dockser, Jeff Fisher, James Kuo, Norm Salmon, Dave Blair, Scott Hankinson, Ken Sghia-Hughes, John Rogers, Michael Jones, Vasilios Brachos, and Eric Costa. To all of them I express my thanks for their cooperation. And to John and Patricia Worden, James Worden's parents, as well.

Solectria allies and partners who made important contributions to my research were Sheila Lynch of the Northeast Alternative Vehicle Consortium, Nancy Hazard of the Northeast Sustainable Energy Association, Boston Edison's James Hogarth, Dave Dilts, and Don Walsh, DARPA's John Gully, investor Jay Harris, Dr. Rob Wills, Richard Watts and Tom Horn of EVermont, and Lin Higley of Ovonic Batteries.

I also wish to thank numerous others for their input and time. They include Jim Ellis and Sean McNamara from GM, SAFT's Jean-Pierre Cornu, U.S. Electricar's John Dabels, CALSTART's Mike Gage and Bill Van Amburg, and SMUD's Ruth MacDougall and Dweight McCurdy. Other contributors include former CARB chair, Jananne Sharpless, and spokesman, Bill Sessa; Catherine Anderson, Paul B. MacCready, Jr., and Alec Brook of AeroVironment; Howard Wilson from Hughes Aircraft Company; Daniel Greenbaum and Dr. Jane Warren from the Health Effects Institute; Jeffrey Bentley from Arthur D. Little; James Merritt and Bill Siegel from the U.S. Department of Energy; Massachusetts' environmental secretary Trudy Coxe and her general counsel Jamy Buchanan; NESCAUM's Michael Bradley and Jason Grumet; Rusty Russell of the Conservation Law Foundation; Chris J. Calwell and Veronica Kun of the Natural Resources Defense Council; along with Ed Bernardon, Noel Perrin,

Dave Maass, David Cohen, Jerry S. McAlwee, Sheldon Weinig, Richard Poirot, Richard Brooks, Jeff Alson, and David Swan. I benefited from the published work of Matthew Wald of the *New York Times*, Daniel Sperling, director of the Institute of Transportation Studies at UC/Davis, and Paul Miller of the Alton Jones Foundation.

For providing me the opportunity to write this book, which took four years to complete, I am grateful to my editor, Herb Addison, who always was encouraging and supportive throughout. Thanks also to Steve Spoonamore and Jeri Theriault for their input on the manuscript in various drafts.

CHARGING AHEAD

1

The Freewheeling James Worden Wins Another Race

C. Michael Lewis

At the American Solar Cup held in Visalia, California, 1988, James Worden set a speed record for a solar car, going ninety miles per hour in Solectria V; here a California State Trooper feigns giving Worden a speeding ticket.

No one can make history who is not willing to risk everything for it: to carry the experiment with his own life through to the bitter end.

—C. G. Jung

James D. Worden and Anita Rajan founded Solectria Corporation in 1989, shortly after he had graduated from the Massachusetts Institute of Technology with a degree in mechanical engineering. By 1994, Solectria had emerged as a technology leader in the electric vehicle (EV) industry, an industry that hoped to reinvent the car. In those days, close observers of the new and speculative industry often referred to Worden simply as "James," confident that the narrow circle of those who knew about electric cars and their technologies would know who they meant. They said things like "James is a genius," "James is the Bill Gates of EVs," and "James is a nerd." All were true, in a sense. But as the CEO of a small, agile company trying to change the biggest industry in the world by putting the most efficient car possible on the roads, he was a little more, too.

Most crucially, James Worden seemed to manifest an inner truth, a deep-seated conviction that what he was doing was right and would prevail. He also possessed that rarest of qualities: confidence before great uncertainty, which makes great achievements possible. Thin, with angular features and dark, impenetrable eyes (he could see out, but you couldn't see in), Worden wore his hair short. He preferred T-shirts and shorts to suits and ties, sneakers to dress shoes. He wasn't, by any gauge, a slacker, and it irritated him to be lumped together with his peers under the Generation-X banner. He loved to race.

Weeks before a race, Worden and his Solectria teammates hardly slept. They pushed themselves to get ready right up to the last minute. They arrived at most races sleep deprived and keen to compete. With Worden driving, they often won: successive American Tours de Sol, the Arizona Public Service Solar Electric 500 in Phoenix, enduros that showed the public electric cars could go farther and faster than critics claimed. Yet Worden didn't seem obsessed by winning.

For instance, on the final day of the 1994 American Tour de Sol, the alternative-vehicle industry race held every May in the Northeast, there was less than a minute remaining before the start and he was nowhere to be seen. Yet there sat his car, in first place. It was a Solectria Force RS, a Geo Metro converted to an electric car: no gasoline engine, no gas tank, no emission controls. The Force RS was supposed to lead the fifty-vehicle field out of Boyertown, Pennsylvania, and head for Independence Square in downtown Philadelphia. But the lead driver, a twenty-seven-year-old CEO in shorts and sneakers, wasn't in his car.

"Where's Worden?" snarled Lin Higley. He was talking to the Force RS's navigator, a nebbish-looking kid with pale skin. Befuddled, the kid, who happened to be James Worden's younger brother, Peter, said he didn't know.

Higley glanced around. He worked for Ovonics, the manufacturer of the advanced nickel-metal hydride batteries powering the Force RS. Higley was the vehicle energy manager, the equivalent of a fuel manager in a conventional car race. Before the Tour de Sol, he'd spent three days at Solectria's shop in Wilmington, Massachusetts, and had come away both impressed with how much work got done in a short time and amazed at Worden's leadership. Now, though, Higley was sweating bullets; he felt absolutely on pins and needles. Two days earlier, on the third leg of the tour, the experimental batteries had juiced the Force RS 214 miles, setting a new distance record for a single charge. "I had one hell of a time keeping up with James," Higley had told several people. "That car handles like there's no tomorrow."

Today, though, was today. The Force RS needed a driver. So where was Worden, Higley wondered as he scanned the motley field. And motley was definitely the right word. There were a few conventional-looking electric cars bumper to bumper behind the Force RS. But then came the solar cars, shimmering eye-catchers with pointed snouts, wings, and insectlike bodies on wheels. Only minutes ago, several solar car crews had been scolding passersby, waving them and their long morning shadows around tilted solar panels. Each solar car needed every precious mote of power its photovoltaic array could capture from the sun and store in batteries. Though the batteries could be charged from the electricity grid, with its wires typically leading to fossil-fuel fired power plants, the rules of the tour forbade that.

Much closer to Higley was James Worden's closest competitor, a red Ford Ecostar. It glowed as though its crew had waxed it overnight. Maybe they had. Ford was the sole entry from Detroit's Big 3: Ford, General Motors, and Chrysler. Ford had sent along an educational support team to explain to the public how the Ecostar worked, and a video crew that broadcast day-to-day highlights on the Ford media network.

Elsewhere across the dewy grass, eager to be off beneath the crisp blue sky on this, a perfect day for solar cars, were vehicles that high school and college teams had built with new and scrounged components, with steel and carbon fiber bodies, with sweat equity and cold cash. Some were glued, others screwed, a few even duct-taped together. And there, in the distance, jawboning with a driver and copilot of one of the cars, was James Worden.

"Get his ass over here!" Higley barked at the now anxious Peter Worden in the Force RS. "Run, run!"

Peter jumped out and ran.

Soon James and Peter Worden were both running back toward their car. They slid inside, the starter finished the countdown, and the Force RS rolled silently out of the display paddock and into the streets of Boyertown.

The route was forty miles over mostly rural roads and was blazed with arrows affixed to signs, sticks, and fences. Nancy Hazard, who worked for the Northeast Sustainable Energy Association in Greenfield, Massachusetts, directed everything—funding, planning, logistics, promotion—as she had for the first five tours. The first NESEA American Tour de Sol, back in 1989, had drawn only four entries. Last year's had attracted thirty-eight. This year sixty vehicles entered, including three motorcycles. In New York City, where the tour had started, each vehicle had passed safety and performance tests. Ultimately, though, this was a road race, with tunnels, traffic, and unpredictable factors. "You got to be a blood-and-guts driver to be in this thing," said Art Liskowsky, a high school vocational skills teacher who had come along with the Enosburg Falls, Vermont, team and its two race cars.

Today, driving through suburbs and then green Pennsylvania countryside that simply merged into the metropolis of Philadelphia, everything went so smoothly for Worden in his Force RS that he neared the finish line long before he wanted to cross it. So he pulled over. The rules of the tour penalized a car for finishing early; that meant it was speeding. Overall scoring was based on efficiency. Daily, each vehicle accumulated tour miles based on the distance driven, if done in a prescribed time window, minus penalties for violations like speeding or running stop signs, if the driver got caught by a tour official posted en route. Some days extra laps were also driven, adding to a vehicle's total score. Overall efficiency and driving range determined the winners in five categories: production vehicles, solar racers, vehicles running on new technologies, like flywheels and fuel cells, mass-transit buses, and an open/hybrid class.

Assured by his performance of winning the commuter class, Worden sat in his Force RS, biding his time and talking with his brother. The car didn't idle because electric vehicles shut down when they're still. It didn't make any noise. It just sat there, clean and quiet.

Other vehicles pulled in alongside it: the Ford Ecostar with advanced sodium-sulfur batteries, which were a promising but hard-to-control energy technology because they operated at 600 degrees Fahrenheit; the MIT Aztec, with its Astropower polycrystalline silicon photovoltaic array (the college team's car, a carbon fiber shell over a light steel frame, weighed only 800 pounds and was shaped like a fish to enhance aerodynamics); and a white Solectria E-10 pickup driven by Sheila Lynch, executive director of the Northeast Alternative Vehicle Consortium (NAVC). The consortium was an alliance of public utilities, environmental groups, high-tech firms, universities, and others eager to get in on the ground floor of a new industry, one that might change the face of America's transportation system and create jobs, if—and this was a big if—the auto and oil industries, together with the Clinton administration, didn't kill it first, according to Lynch.

Lynch, an environmentalist with a degree from Harvard, admired fellow Bostonian James Worden. She'd been pivotal in seeing that Boston Edison Company, an electric utility, and the Pentagon, through its Defense Advanced Research Projects Agency (DARPA), had invested major dollars in Solectria's Sunrise project.

Indeed, the Sunrise project was one reason James Worden probably shouldn't even have been racing this year. Racing's advantages—setting deadlines, making decisions under pressure, putting one's pride on the line—had shaped both Worden's character and Solectria's growth. But racing also consumed lots of time. Worden's love of racing typically disrupted Solectria for a week before an event, as everyone worked to get ready, and for days afterward, as everyone recovered. Repeated disruptions were tough on a small company, especially one with big dreams to build the best electric cars in the world. The time could be better spent, Worden himself acknowledged, on projects in the shop, the most important of which was the Sunrise.

Sunrise was Worden's dream car. It was a from-the-ground-up electric car that could be made at low cost and powered by conventional batteries. When more advanced batteries came onto the market, like the Ovonics nickel-metal hydrides in the Force RS, they would improve the car's performance. Presently, back in Wilmington, a team of four guys (their number would grow to twelve over the next few months) was designing the first Sunrise prototype, a two-door, four-passenger sedan made from lightweight composites rather than traditional steel. The question was, could a small, audacious company of mostly twentysomething tech-nodes blow open the door to a new way of making a car? The answer was, of course not. Not with the Big 3 and the oil industry spending millions to cripple the momentum that was pushing Sunrise into being. But no matter. This was America. There was no law against thinking the impossible dream—just unwritten laws making such a dream patently impossible.

Unwritten law 1: Only large automakers, the so-called "Big Boys," build cars.

Unwritten law 2: Mandates helping little guys like Solectria build clean cars would get killed—by the Big Boys.

Unwritten law 3: America's institutions were too outdated to do much about unwritten laws 1 and 2.

Today, though, more positive concerns were occupying Worden, like winning another race. And it was time to cross the finish line.

Worden had planned to be first. It was important because the press perceived the winner to be the driver who was first across the line. Today, though, Helios the Heron, a car made and now pushed by its team of Vermont junior high school kids and their advisers, rolled into sunlit Independence Square first. Second, Worden accepted the faux pas gracefully. Other vehicles followed. Race team crews, TV crews, and fans, including school kids carrying miniature solar cars they'd race later on the sidewalk, greeted the vehicles with applause and cheers. Now it was official: Worden's Solectria Force RS had won the American commuter class, Ford's Ecostar had won the production class, and the MIT Aztec had won the solar commuter class of the sixth running of the NESEA American Tour de Sol.

2

Then and Now: Rebirth of a Dormant Industry

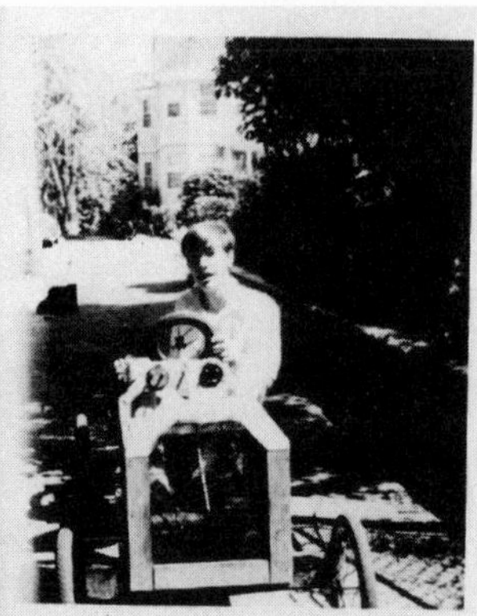

Mr. & Mrs. John Worden

James Worden driving Solectria I, which won first prize in the Massachusetts State Science Fair in 1984, down his parents' driveway in Arlington. On the right he's at the wheel of an earlier experimental vehicle that was modeled after an early 1900s racer. Worden originally built this race car as a gasoline model, but later, in 1982, when he was in eighth grade, converted it into an electric.

Automobiles are good things and some of the people who own them or drive them are fit to be trusted with them; but to tell the thing exactly as it is, most of them are not.

—*Manchester Union,* July 21, 1905

Fast reverse: eighty-nine years ago: the first Glidden Tour: New York City to Bretton Woods and return.

It was July 1905. Charles Jasper Glidden, an adventurer, was inaugurating his famous race. Glidden had been born and raised in Lowell, Massachusetts, thirty miles north of James Worden's hometown of Arlington. Like Worden, Glidden had found his direction in life while young. Working as a telegraph operator in Lowell, Glidden befriended Alexander Graham Bell, an inventor seeking subscribers for his newfangled idea, the telephone. Twenty-year-old Jasper Glidden, convinced some people would pay to talk to their neighbors through a wire, soon signed up fifty takers in Lowell.

Ninety years later such risk takers, willing and often eager to take a chance with something new, were called "early adopters." Early adopters had bought the first VCRs in the 1970s, the first PCs in the 1980s, and, if electric cars were to succeed in the 1990s and beyond, they'd be the first to own them.

Back in 1905, thirty-three early adopters of the automobile, or "misguided cranks," as critics sometimes called them, entered the first Glidden Tour. Some of the drivers were destined for the history books, including R. E. Olds who drove a Reo; Percy Pierce, who drove a Pierce; Benjamin Briscoe in his Maxwell-Briscoe; and A. L. Pope in a Pope-Toledo. After leaving New York, the racers sped northward on mostly dirt roads. The rules were simple: no observers, no timing, no limit on spare parts. In addition to the cars, three motorcycles, four steam cars, and two trucks entered. Engines ranged in size from six to sixty-four horsepower. The inaugural Glidden Tour was an 871-mile jaunt through the hinterland. The object was to show Americans that cars weren't "rich men's toys," but instead were meant for traveling. The tour also promoted better roads.

The Glidden Tours, which continued through 1913, and the NESEA American Tours de Sol of the 1990s were similar in some ways, quite different in others. Both showcased new technologies. Entrants were on the cutting edge of what they were convinced was the new, and their goals were often obscure to their peers. The cars of neither era made good sense to the man in the street, one enamored with horses, the other with horsepower.

In 1905, though, the first generation of electric cars competed strongly with their gasoline-powered counterparts. Their key components, batteries and electric motors, had reached a high level of performance. Batteries, which had been available since the 1850s, allowed electrochemical power to be stored, discharged, and recharged. Electric motors had reached a certain degree of sophistication since the first one had been built in the 1830s by a Vermont blacksmith named Thomas Davenport (Davenport used it to run drills, lathes, and a printing press, as well as a miniature train; ultimately he was far ahead of his time, but unable to find backers for his invention, he died penniless). Even regenerative braking, which recharged a car's batteries while it went downhill because currents can be reversed, had received a U.S. patent in 1888. So, by the 1890s, electric car components were at the stage where the inventive could build rudimentary vehicles and drive them out into the streets.

What was missing was a reliable way to keep the cars running, that is, a charging system. This crucial enabling technology was in its infancy in cities; out in the country, it was virtually nonexistent. Comparatively, the enabling technology for gasoline vehicles, gasoline stations, increased in number because of the ease of installing a storage tank by a grocery or hardware store. All a gas car owner had to do was pull in and fill up.

Comparisons between gas and electric cars in the early years of the automotive era tended to emphasize that electrics were quieter, cooler, less prone to breakdowns, and ideal for lady drivers, since there was no need to crank them over to start the motor. Gasoline cars, on the other hand, had greater range, accelerated faster, climbed hills better, and were ideal for touring, a favorite pastime of men. This gender gap—men wanting speed and distance, women wanting quiet and ease of operation—played a role in the ascent of gas cars. In the predominantly patriarchal society of the early 1900s, in which families owned one car, men decided what the car would be. Typically, notes Michael Brian Schiffer in *Taking Charge: The Electric Automobile in America*, "the struggle between the sexes ended in the purchase of a gasoline car."

The growing popularity of gas cars mirrored another truth about early-twentieth-century America. People were more attuned to mechanical technologies than to electric ones. Electricity as an energy source was a relatively new phenomenon. Mechanical awareness, in contrast, was more highly developed, with water power and geared machinery operating across the nation. Nearly a century later, quite the opposite was true; electrical devices mirrored the times, from video games for kids to computers for adults to chip-operated robots on assembly lines. Most late-twentieth-century Americans knew more about bytes than they did about torque.

At any rate, gas cars got bigger, faster, more refined, and, once mass production of Ford Model Ts began in 1910, much cheaper. The result was that the electric car, its manufacturers and service outlets, its enthusiasts and female devotees gradually shrank in number. By the 1920s, the electric car had pretty much disappeared from the American scene.

Eighty years later, no company had yet resurrected the long-defunct electric car industry. But there was no lack of potential pretenders to the throne, in America, Japan, or Europe.

For instance, at the Thirtieth Tokyo Motor Show in November 1993, six months before the Sixth American Tour de Sol, more than a dozen zero-emission electric and ultra-low-emission hybrid-electric vehicles that blended gasoline and electric car technologies were on display. They included the Mitsubishi ESR, or Ecological Science Research-mobile, with its bug-smooth blue skin and solar-panel roof to keep the interior cool while it was parked; the Toyota AP-X, which promised to combine environmental friendliness with sheer driving pleasure; and the cute, pugnacious BMW E1, a hybrid-electric car whose power plant gave a driver three options: "zero emission" mode for in the city, ICE mode ("ICE" stands for the old standard, the internal combustion engine) out in the country, and hybrid mode, which integrated both electric and gasoline power to increase driver options, lower emissions, and maximize efficiency.

The theme of the Tokyo Motor Show was "Ecolution in Car Innovation." The word "ecolution" was a hybrid, an etymological marriage of "ecology" and "evolution." The prototype vehicles caught the eyes of some of the 100,000 Japanese and foreigners who paraded daily through the massive Nippon Convention Center.

None of the concept cars were yet on a manufacturing track. Basically, the automakers were hedging their bets. The old standard, the ICE vehicle, still held center stage, with sport utilities promoted the most zealously. For example, Daihatsu had a wall-size video screen behind its MP-4, a small sport utility. On the screen a clip repeatedly showed a happy young Japanese couple driving the MP-4 along a deserted beach, the surf white, the couple ecstatic, no other humans in sight. To either side of the MP-4 on display in front of the one on the screen, hostesses wearing black shorts and boots and red or yellow jackets danced around the little surfer RV in an ICE-age tribal number. Gawking Japanese boys photographed the dancers. Over the sound system the surf boomed. One had to wonder, how far from Tokyo do you have to go to find an empty beach to shoot such a commercial?

Despite the emphasis on ICEs and no commitment from any major automaker at the show to mass-produce EVs in the near future, it seemed only a matter of time. Robert Stempel, former CEO of General Motors, said as much. In America Stempel wanted to start an electric car company. The ICE had passed its peak efficiency, Stempel said. He praised the ICE for its great progress, but the costs, lifetimes, and reliability of ICEs were all being squeezed by expensive, complicated new technologies in order to reduce pollution and improve mileage. Electric motors, in contrast, were on a steep improvement curve. They had far fewer parts, ran emission-free, and lasted longer. The ICE was not at a total technological standstill, Stempel said, but improvements inched along incrementally, at high costs for small gains.

It is interesting to recall that, back in 1905, gasoline had not been available on every other street corner in America. Kerosene, or "new light," as people called it, was; you found new light in drugstores, at grocers. Gasoline, a by-product of the refining process, ranked much lower in economic importance than kerosene, fuel oil, lubricants, and products like petroleum jelly and paraffin used in canning food and making candles. Some oil refineries dumped excess gasoline into rivers at night to get rid of the stuff. In order not to run out of fuel in that era, drivers strapped jerry cans of gasoline aboard the Packards, Peerlesses, Darracqs, and other vehicles on the Glidden Tour. In the 1990s, a portable generator trundled along with the American Tour de Sol. It had to plug into the enabling technology, the overhead electricity grid, at each stop. The charging needs of EVs on the open road were at least as challenging in the 1990s as finding gasoline had been in 1905. A series of easily accessible, fast-charging stations along roadsides and at major intersections, or as integral parts of the typical gasoline station, would ease the market penetration of EVs in the 1990s and beyond. But easing market penetration for electrics was not something either oil or auto companies were eager to do.

Similarities between the American Tour de Sol and the Glidden Tour included the fact that both tours ran in traffic on real roads, not on test tracks; also, both tours were nuisances to those enamored with the status quo: horse and buggies in 1905 and gasoline cars in 1994. An "unmitigated nuisance" one newspaper editorial labeled the goggle- and glove-wearing, speeding motorists slashing across New Hampshire in 1905. New Hampshire constables, exasperated because their horses couldn't catch the speeders, even strung ropes across roads to net the cars. Normal folks, traveling at a pokey gait in their buggies or broughams, felt menaced "for no earthly reason other than to afford amusement to a lot of strangers," wrote the *Manchester Union*. "We say it is an outrage, and if these people think of coming here another year, we hope the law against speeding and scorching will be promptly and vigorously enforced. . . . Let a few of them stay in jail for two or three days."

Internal combustion engine racers then, EV racers now, helped each other with repairs, advice, and encouragement, with rare exception. Paper streamers marked the routes for both tours. But in 1994 the papers were arrows affixed to objects. By comparison, in the 1907 Glidden Tour a reporter in an Aerocar, one of the first vehicles with an air-cooled engine, described the pace car "in the absence of signposts, scattering paper along the road and particularly at intersections, where doubt existed as to the correct route." He added that the Aerocar ran well but kept getting flats. Soon he was helping the exasperated driver stuff the tires with grass "until the rear axle protested and left us completely stranded." Unfazed, the reporter said he stuck out his thumb and hitchhiked.

Hotels were the preferred accommodations of old. Room rates shot up when the Glidden Tour passed through town. Not that it mattered. Though muddy, sweaty, and tired, many of the competitors were already millionaires. Less well-heeled Tour de Sol competitors often pitched tents, slept in recreation vehicles, or went looking for an inexpensive motel.

Drivers in both tours were tough. Skids across wet roads, fixing wheels in the rain, devising makeshift steering mechanisms when the regular ones malfunctioned—these were commonalities drivers and teams dealt with. One notable difference was their respective attitudes toward publicity. In the 1990s, the small car companies hungered for media coverage, which gave them credibility and leverage with investors; in the Victorian era, gentlemen tried to stay out of the papers, even with their entrepreneurial endeavors.

Safety-wise, the Glidden Tours, which ran for many more miles, were more dangerous than the Tours de Sol. "The Blood-Red Trail of the Gliddenites," one paper called the race in 1907. That year the driver of an open Packard died from internal injuries after his car rolled. If a driver were to die on the American Tour de Sol, media attention on a burning electric car or a flattened solar racer might send an image, and a message, around the nation that the cars were hazardous, thus derailing the entire industry. A century ago, deaths and injuries were chalked up as part of the learning curve, as costs of automotive progress in an ICE age. Entrants agreed to take responsibility for the risks.

For the Glidden Tour, the premier reliability test for cars of its time, competitiveness killed the fun by 1910 or so. Winning became everything. As the auto industry boomed, manufacturers felt they had to have that trophy. Contentiousness, cheating, even lawsuits, hounded the event its last few years. The tour ended in 1913, a victim of its own importance. The American Tour de Sol, in large part because it was not yet important to major manufacturers, had avoided the competitive ugliness that killed its predecessor.

All in all, the two tours complemented each other in a pattern that connected the past with the present. The questions in the 1990s were these: Were electric cars reinventing the automobile industry as the gasoline car had invented it a century before? Were ICE vehicles doomed? Were they energy gluttons crowding the highways during the sunset years of oil and the sunrise years of global warming? Was the ICE age an aberration? Was the twentieth century a mere sidebar in the long story of mobility and its energy sources, a peppy, speedy run that had colorful moments but wouldn't mean much by 2020 or 2030 when other energy sources ruled the marketplace?

3

James Testifies on the Promise of Clean Cars

C. Michael Lewis

Catherine Anderson, the mechanic, holds up the cockpit glass of MIT's Solectria V before the American Solar Cup, Visalia, California, 1988. In the race, Worden drove the car almost ninety miles per hour across a long, flat stretch, setting a new solar car speed record.

Innovative thinking, a willingness to take technology risks, and a strong commitment to research and engineering can overcome many obstacles.

—Testimony by James Worden before House Science, Space and Technology Committee, June 1994

James Worden thought so. "We've pulled these cars out of the hands of the backyard hobbyists and put them in the hands of people," he said, not long after the 1994 American Tour de Sol. "They're usable. They're tinker-free. It used to be, definitely, one of those things—like with gas cars in 1898 or something—if you were willing to tinker and get into it, you could have one. Until pretty recently, electric cars were reserved for people who didn't mind playing with them. Now we've made them simple and very efficient, so they can go far enough without trouble."

Worden had said as much before the Subcommittee on Energy of the House Science, Space and Technology Committee in Washington just weeks after the Tour de Sol. Wearing a suit and tie, he told the congressmen that he strongly disagreed with the Big 3's contention (a contention that would gradually change) that EVs would always be expensive. There was nothing inherent in an electric car that should make it higher priced than a gasoline one, he said, yet electrics were different. The heart of an EV was electronic, not mechanical, he explained. Instead of hundreds of precision-engineered moving parts operating at high temperature, there was a motor with one moving part and a controller with no moving parts. An EV industry would need a new set of suppliers, and perhaps a different assembly process. Were the Big 3 uniquely qualified to make EVs?

"Perhaps," Worden said. "However, perhaps not."

The Subcommittee on Energy was especially interested in the state of battery developments. The federal government was investing hundreds of millions of dollars in what some critics called a crummy energy storage device. Although Solectria was not a battery company, Worden had experimented with many types of batteries over the years, from conventional lead-acid to advanced nickel-metal hydride, from dumb to smart, from benign to dangerous. Worden's testimony was ambivalent about the future role of batteries. They might never provide the performance and range of gasoline, he conceded, but that wasn't the important issue. Other technologies, such as fuel cells, flywheels, and supercapacitors, all held out promise to eliminate batteries.

Meanwhile, Solectria was not waiting for some big battery breakthrough, some "superjuicer" whose energy density was close to that of gasoline. Typical city driving didn't need a superjuicer, Worden said. Maybe lightweight composites were just as important as batteries, he said. Composites lowered the weight of a car so it needed less energy to go the same distance.

Toward the end of his testimony, Worden trundled out an often heard comparison between the emerging EV industry and the computer industry of a decade earlier. He said, "In a decade, Microsoft Corporation grew from a start-up to one of America's most successful companies by beating more entrenched competitors in a rapidly changing market. In a fast-moving business environment, smaller entrepreneurial companies are able to innovate faster and take greater technological risk. The EV industry may be the same."

Not long after Worden's testimony before the Subcommittee on Energy, *Business Week* put EVs on its cover. Large black letters said "ELECTRIC CARS," and small white letters across the hood of a Ford Ecostar asked, "Are They the Future?" A reporter drove an Ecostar from smoggy Los Angeles to Sacramento, a distance of more than 500 miles, in five days "to find out the truth about those four-wheeled hot potatoes that environmentalists love and Detroit hates." He concluded that EVs would make it—somehow. Government incentives, tax breaks, and free parking spaces would help jump-start the industry, which EV supporters had been saying for years. "The drive to build a worthy electric vehicle is every bit as challenging as the Apollo moon shots," the reporter said. "And more useful on earth."

Useful on earth or not, to date no automaker had committed to mass-producing an electric car. Yet the pressure was on. In May 1994 the California Air Resources Board, an appointed regulatory board responsible for the state's air quality, had reaffirmed its commitment to a zero-emission-vehicle mandate beginning in 1998. According to the mandate, which was a key piece of California's latest low-emission-vehicle plan, in 1998 2 percent of the vehicles sold in California by the Big 7 (Ford, Chrysler, General Motors, Honda, Toyota, Mazda, and Nissan) had to be emission-free. In 2001, 5 percent had to be emission-free, and in 2003, 10 percent. That meant, with present technologies, battery-powered electric cars. Despite the potential market created by the mandate, the Big 3 were trying to kill it, together with similar ones in New York and Massachusetts, instead of gearing up to make big strides in improving component and charging technologies.

At the heart of the Big 3's objection to electric cars was the argument that, mandate or no mandate, Americans wouldn't buy electric cars yet because they performed poorly and cost too much. Yet the automakers were going to have to spend hundreds of millions of dollars to make them. To undertake such a marketing effort before the technologies were better, and batteries more advanced, would result in higher prices on gasoline cars and would even turn off EV buyers because they would be disappointed. This would move the whole clean-car movement back instead of forward. Who would pay? The consumer, the automakers claimed. "When people become aware of the hoax that's been perpetrated," said David Cole, director of the Office for the Study of Automotive Transportation at the University of Michigan, "they will become unglued. At some point, there will be hell to pay."

The ambivalent attitude the Big 3 brought to EVs, and their dislike of mandates forcing new technology into the marketplace, combined to create an unusual opportunity for Solectria and other small, agile companies willing to take big risks. The rewards for success could be big. Automobile manufacturing and related businesses employed one out of seven working Americans and, in some European countries, one out of five. But even in their wildest dreams, there was no way Solectria and the other upstart EV companies could supply California's projected forty thousand electric vehicles for 1998, never mind additional thousands in the Northeast, if the mandates held there.

Something had to give—either the Big 3 or the mandates.

4

An Idea Whose Time Has Come?

Dennis Renault

The California Air Resources Board passed its zero-emission vehicle mandate in late 1990; the mandate was battered relentlessly by auto and oil interests, as depicted here in a cartoon published in the *Sacramento Bee* in 1994.

During the past three decades . . . new generations of Americans with social needs, attitudes, and lifestyle expectations have reached car buying age, many of these people are apt to regard the electric city car as something quite desirable—at least in principle.

—Michael Brian Schiffer, *Taking Charge: The Electric Automobile in America*

On the surface it was hard to fathom why the Big 3, with total profits of $13.9 billion in 1994, a workforce of around 600,000 and branches linking their Detroit-based headquarters with dozens of foreign countries, were worried about the struggling EV industry. In 1994, that industry's four largest independent companies—Solectria in Massachusetts, Renaissance Cars in Florida, Unique Mobility in Colorado, and U.S. Electricar in California—had losses estimated at several million dollars, about 400 employees, and a few agreements with companies and governments in Europe and on the Pacific Rim. The facile comparison between the American computer industry of the 1980s and the automotive one in the 1990s, which James Worden mentioned before the congressional oversight committee, was a turn-on to EV advocates because it showed quick, small companies outflanking large, slumbering ones. But it was a stretch. No small EV company was positioned as solidly as Apple and Microsoft—the two computer industry companies often referred to—had been in relation to their competition. In addition, the Big 3 were definitely in the EV game, albeit complaining all the time about the rules. They were spending tens of millions of dollars on

EVs while simultaneously attacking their performance and the mandates with tens of millions more. No one familiar with recent automotive history was surprised by the seemingly contradictory strategies. When the subject was regulations or mandates, Ford, General Motors, and Chrysler continued to operate as they had for decades: denying, obfuscating, disseminating disinformation along with good information, challenging, litigating, and finally—if all else failed—doing what was asked of them in the beginning, with concessions. The concessions were the goal of all the strategic, and expensive, foot-dragging in order to guarantee least feasible retreat from the status quo.

The contention that small, entrepreneurial companies could innovate faster, and take bigger risks, in the giant-dominated world of automobiles had yet to be proven, too. It sounded good. It had even worked in some industries, such as personal computers. But like a small plane flying low, little companies could also make hasty, foolish mistakes and take a nosedive.

Yet the Big 3 did have cause to worry. There is nothing more powerful than an idea whose time has come. Air quality regulators, politicians, environmentalists, and even car buyers, according to some polls, wanted to see clean cars on America's roads. They would help reduce air pollution, decrease oil imports, and create new jobs, proponents said. Clean cars might help reverse global warming and usher in the dawn of a new energy consciousness, a consciousness some critics argued was necessary if the human race was to sustain itself.

At the top of GM, Ford, and Chrysler, executives knew all this. So the thought that the electric car might be the next idea whose time—if it hadn't exactly come—loomed on the near horizon, caused considerable consternation and soul-searching. But the real issues were traditional ones: power and money. The Big 3, and their competitors, had huge capital investments in plants, processes, and technologies making ICE-age cars. EVs would demand radical change: new technologies, manufacturing processes, fueling infrastructures, and repair networks. Besides, in the early 1990s the Big 3 were just emerging from a tough decade during which Chrysler could hardly afford to keep the lights on, GM laid off a quarter of a million people, and Ford had revived after a close call with bankruptcy. The Big 3 were again making money, getting back to what automotive analysts called "trend volume" and enjoying a few good years. The unspoken attitude was, why mess around with electric cars when the world's appetite for ICEs was on the upswing? China, India, Brazil, South America—all were promising markets for petroleum-powered vehicles.

That made sense. Deep in the automotive culture there was a taproot sunk into petroleum. Pulling that root out, figuratively as well as literally, was going to resemble the challenge King Arthur had faced getting the sword Excalibur out of stone. Something almost magical was going to have to occur if the American automobile industry was ever going to wean itself from the fuel that gave it birth. And if American drivers were going to recognize and deal with their addiction to gasoline-fueled mobility at a supported price, the biggest support being the cost of military preparedness to go anywhere in the world to guarantee an uninterrupted flow of oil.

In 1994, which way the Big 3 was going to shift next in its attitude toward clean cars was particularly confusing. The companies worried about the rebirth of what was thought to have been a dead industry but now showed it had only been dormant, lying there since the turn of the last century, when EVs briefly had competed with ICE vehicles for dominion of the roads, waiting to be recharged.

One reason Solectria and the other small EV companies were important was that they showed the potential of the EV revival. They put new technologies in cars first. They put drivers behind the wheel at public demonstrations, letting them decide for themselves how the cars performed. The small companies kept the pressure on the Big Boys to change, to stop bad-mouthing the new, to stop saying EVs performed poorly. The small companies gave heart to regulators and to the rare politician willing to publicly endorse the idea of getting EVs in the marketplace soon.

Of course, the small companies couldn't supply large numbers of vehicles. So, in truth, they were less a threat to the large automakers than they were an embarrassment. If EVs really were to make it into the mainstream, big automakers had to commit to making them.

5

James's Second Car: A Goofy-Looking Aluminum Thing

Courtesy of DOW Chemical Company

Solectria II, built by Worden during his senior year in high school, was a primitive vehicle compared with Solectria V, shown here with James leaning over the cockpit after the American Solar Challenge in 1988. Wedge shaped like Solectria I, with an aluminum skin, three wheels, and a sloping nose of solar panels, Solectria II was driven thousands of miles by Worden.

James was this wacky kid who knew absolutely nothing about machines, or machine design. All he knew was that electric cars were cool, and they had to be light.

—Ed Trembly,
Solectria's first full-time employee

James Worden had built his first roadworthy electric car when he was still in high school. He called it Solectria I. The paint was still drying on Solectria I as James and his father rushed it to an exhibition hall, where it won first prize in the 1984 Massachusetts state science fair. The next year, as a high school senior, Worden built Solectria II. Ed Trembly, who later became Solectria Corporation's first full-time employee, remembered Worden bringing a piece of the car into a machine shop in Arlington. It was a lazy Susan, the revolving part of a table. Where Worden scrounged it he didn't say, but he told Trembly he was going to attach the front wheel of his car to the lazy Susan.

Recalling the incident years later, Trembly chuckled. "He wanted to make this electric car," he said, "this goofy-looking aluminum thing. He was this wacky kid who knew absolutely nothing about machines or machine design. All he knew was that electric cars were cool, and they had to be light. I had visions of him ripping his steering wheel off on the first pothole. After much wrangling, I convinced him to put a bolt through the middle of it to hold it together if the lazy Susan failed. But he drove that car. James must have put 10,000 miles on that car. He drove it to MIT for four years. The front wheel never did fall off."

During his freshman year at MIT, Worden went to Mirak Chevrolet in his hometown of Arlington and asked the dealer to sponsor his next car. He got five thousand dollars, which he used to build Solectria III, a solar-powered electric racer, mostly at his parents' home on Jason Street. He entered Solectria III in the 1986 Swiss Tour de Sol.

Predecessor to the American Tour de Sol, the Swiss race had originated in 1982. It was romantic, ambitious, and picturesque. For six days and more than seven hundred miles, competitors drove their vehicles through the Alps: over high passes, along older roads, through villages. The teams came mostly from Germany and Switzerland; they had university and corporate sponsors, and they wanted to push alternative technologies into the automotive mainstream. The Swiss Tour de Sol radiated a sense of serious play, of the future in the making.

Under optimum conditions—bright sunlight and cool air, which are ideal for photovoltaic cells to convert sunlight into energy—the first-rate solar cars accelerated to sixty miles per hour in less than ten seconds and cruised at highway speeds. Lightweight and eye-catching, many of them relied more on recent advances in bicycle technology—brake components, light metals, composites, wheel designs—than on automotive advances. This repeated an old pattern. In the early twentieth century new bicycle technologies, like the steel-tube frame, chain drive, and the pneumatic tire, which John B. Dunlop invented in Ireland in 1888, had been crucial preludes to automotive advances. Now the solar cars with their modern bicycle components often used less than 10 percent of the energy of a conventional gasoline car. They looked a lot different, too—stranger, more aerodynamic, and more vulnerable. And instead of plugging into the electric grid, they relied on the sun to power their limited number of batteries. The photovoltaic arrays that collected power were expensive and cumbersome, however. Together with complex controls and narrow safety margins, the costly arrays made mid-1980s solar cars unlikely to replace gasoline cars any time soon.

The lack of safety features in James Worden's car worried the nineteen-year-old's parents. Solectria III had a gear that revolved at high speeds only inches from their son's head as he drove in a semi-prone position. "We were scared to death," recalled John L. Worden III, James's dad.

James Worden returned home from Switzerland trophyless but recognized as an unusual nineteen-year-old—even by MIT standards. That fall, back in school, he used the recognition to promote himself and to help establish the MIT Solar Car Racing Team, which Dow Chemical Company agreed to sponsor.

6

Racing Solar-Electric Cars at MIT

C. Michael Lewis

During a break in the competition during the 1989 American Tour de Sol, the MIT team lifts James Worden in front of the New Hampshire state capitol. From left to right: Diane Brancazio, Bruce Larson, Alexandra Worden, Catherine Anderson, Erik Blasch, James Worden, Janey A. Pratt, Anita Rajan Worden, Mark Henault, Andrew Heafitz, Gill A. Pratt, unknown, John L. Worden IV.

Ride on! Rough shod if need be, smooth shod if that will do, but ride on! Ride on over all obstacles, and win the race.

—Charles Dickens, *David Copperfield*

When Worden put up posters inviting students to come to a meeting, he caught the eye of Catherine Anderson. A freshman at MIT, Anderson had also gone to Arlington High School. There, she hadn't known Worden well, though they had played together in the band (she violin, he trumpet), but she had worked some on Solectria III. During her senior year in high school Anderson had been repairing bikes at Bicycle Corner in Arlington when Worden came in and asked for help on the solar race car he was building at his parent's place with the $5,000 he'd received from Mirak Chevrolet. Anderson had made special twenty-inch wheels for the aluminum frame. At MIT, Anderson soon became a core member of Worden's fledgling solar racing team. She was the mechanic.

Five students quickly formed the core of the team: Worden; Anderson; Gill Pratt, who was the electronics expert; Erik Vaaler, an older graduate student who was knowledgeable about lightweight composites and, just as important, about dealing with MIT professors and administrators; and David Brancazio, a mechanical engineering student. Woody Flowers, senior professor of the mechanical engineering department, also played a vital role. Flowers couldn't get official status for the team (someone might get killed while racing cars—not the kind of thing administrators looked at approvingly), and with the status certain benefits and facilities that could be helpful, but his tacit approval of the team's actions and spirit provided something just as important: a well-respected senior professor's faith in the students' judgment of what was worth doing. Flowers also took a special liking to Worden and Anderson.

Not getting official club status at MIT did not bother James Worden. In fact, he saw it as sort of a blessing. Now the team could avoid the school's bureaucracy, move faster and with fewer restrictions, and put all its money into solar race cars rather than giving 40 percent of it to MIT for administration and overhead, which approved clubs were required to do. On the negative side, the team lacked an official adviser. "It would have been tenure suicide," Anderson said. Team members had no sanctioned storage space; dorm rooms, closets, empty spaces under tables had to do. They had no official work space; sometimes work got done on loading docks, in parking lots, in labs that were supposed to be off-limits. And tools were often in short supply, resulting in the misuse of those that were available. Anderson called the team's mode of operation "guerrilla engineering."

One tactic of guerrilla engineering that was absolutely essential was getting lab time. But labs were normally off-limits. The team sneaked in at night. The cryogenics lab, for instance, where professors taught the physics of extremely low temperatures, could be entered through a window in an adjoining women's rest room. MIT's biggest lathes—with four-foot chucks and twenty-foot beds—were in the cryogenics lab; Anderson used them to turn magnesium billets into wheels. In the architecture building, after the professors had packed up and gone home for the day, the team sometimes laid up composite wings, which held photovoltaic arrays, and shells that became car bodies. Working with composites was tricky. Erik Vaaler had some experience with the processes, but actually making a decent shell out of composites, which mix fabrics and resins in matrices with varying drying times, required a fair amount of orchestration and finesse. It wasn't brain surgery, yet resins had varying pot lives, fibers had to be laid up correctly for strength, and a handful of volunteers, with their brushes and cups, had to be given direction. The fumes were also a problem, especially those from polyester resins. They traveled into vent ducts and to distant classrooms, where they were breathed by the occupants the following morning. There were complaints from some of the architecture professors. The result? The team switched to epoxy resins, which were more toxic than the polyesters but emitted very little smell.

One place Worden, Anderson, and the others were welcomed was the MIT student shop, managed by 385-pound Joe Caligiri, or "Tiny." Tiny's shop was an official space for student projects. Few student projects were as big as a car, however, so the solar racing team took up a lot of space. The zealous members pushed even patient Joe Caligiri over the edge on occasion. Once, for example, a composite mold hardened on the cement floor late at night, and no one returned to clean up the mess before dawn. When Caligiri came in, he was livid. Later, Worden, on hands and knees, chipped the mold off the shop floor.

During the three years Worden's team persevered, it experienced occasional crackdowns by MIT authorities. Reprimands were issued, locks to doors changed. The cryogenics lab, Tiny's shop, the artificial intelligence lab (whose milling machines, normally used on robots, were very useful) were doubly off-limits. Appropriate acts of contrition had to be enacted before offended administrators and professors.

"I'd go in if the approach called for, 'Gee, we'll never do that kind of thing again!'" Anderson confessed. Erik Vaaler would acknowledge the complaints with a dismissive "Oh, kids will be kids." Worden, at whom much of the irritation was directed, was not great at apologies; "Oh, I'm really sorry," he'd say if cornered.

But solar race cars were the thing, not apologies for circumventing the status quo at a school known around the world for the audacious and meaningful achievements of its graduates. "There was a real point of pride that we did what we did without faculty support," Anderson said. "James held things together. The rest of us got compassionate and put in hours, but he was really the catalyst. He already had the art of collecting the right people around him."

The right people had brain power, vision, and an understanding of efficiency. They weren't afraid to take chances. In such a group Worden was not a tyrant, but he did possess an iron will.

"The clashes of will were monumental!" Anderson said. Worden disagreed. What Anderson saw as bitter fights he considered only "heated discussions."

The team's goal was to build light, efficient, race-winning vehicles of their own invention. And they did. The results fit into an expanding alternative-vehicle world, a far-flung world of mavericks and tinkerers, of microchips and bicycle parts, of romantic races and practice sessions alongside the Charles River. It was a world far removed in many ways from that of the conventional automotive industry, with its suits and executives, its technical centers and design studios, its vast factories and multi-million-dollar marketing budgets.

7

Air: This Marvelously Tattered Canopy

Dan Foote

This most excellent canopy, the air, look you, this brave o'erhanging firmament, this majestical roof fretted with golden fire, why, it appears no other thing to me but a foul and pestilent congregation of vapors.

—William Shakespeare, *Hamlet*

In 1987 and 1988, environmentalists were already beginning to call the entire 1980s the "lost decade" because of the hostility of the Reagan administration toward the pro-environmental agenda that had been set in the 1970s and the administration's success in stalling that agenda's momentum. Not that James Worden, Catherine Anderson, and the rest of the Solar Car Racing Team cared much about lost environmental initiatives or momentum. Or if they did, they didn't preach about it. It wasn't their style to get boisterous and judgmental about suspected wrongs. Their way of effecting change was more indirect; it was the way of the engineer: testing hypotheses, doing experiments, making stuff and improving it.

"We were more technology junkies than environmental crusaders," Anderson said. "You really have to love the technology to build the kind of vehicles we were building. If your first great love is protecting the environment, you're out there chaining yourself to trees because that's the most direct thing you can do. But if you want to change things in the long term and make technology that improves things further down the road, you're balancing technology and concern for the environment."

Just the same, environmental concerns were strong driving forces behind the small but global search for cleaner cars and their technologies. Of these concerns, the most important was the need for clean air.

"This most excellent canopy, air," as Hamlet put it, sustains life. A thin membrane, it keeps people warm, screens out harmful radiation, and provides oxygen to breathe. At the dawn of the automobile age, in the early 1900s, given the apparent immensity of the sky and the fitfulness of winds, it seemed reasonable to assume that this most excellent canopy, air, would not become "a foul and pestilent congregation of vapors," another phrase from *Hamlet.* Such a transformation of the atmosphere seemed far beyond the puny effusions of man, his industry, and his machines.

Gradually, though, and especially in the 1970s and 1980s, the world learned otherwise. Like rivers, seas, and gradually the oceans, the atmosphere too had become sick in places. Metaphorically, if the atmosphere was like a skin covering the globe, then the skin had cancer and it was spreading.

The most telling symptom was smog. To many Americans, smog was synonymous with air pollution. Of the dozens of tailpipe gases, chemicals, and particles that combine in various ways with industrial and natural emissions, smog was the best known, a unique by-product of the twentieth century, a fascinating, unstable, shifting mirror of the times. In the 1990s, smog, which had actually been around since the start of the industrial age, was temporarily upstaged by new and less predictable contenders, such as excess carbon dioxide, the main contributor to global warming; air toxics, such as the cancer-causing benzene, which you can smell at the gas pump; and micron-sized particulates, which appear as little black clouds erupting from diesel tailpipes and are associated with a lengthening list of lung, heart, and internal diseases. Daily, drivers everywhere helped produce smog in vast quantities. The stuff usually rose with the sun, as drivers started their cars. It crossed city and state boundaries with impunity. It flowed up river systems, clung to shorelines, got trapped by inversions, which acted like lids to hold in rising air currents, and was carried great distances by storms. Once the sun set, smog even got consumed by one of its own components, oxides of nitrogen (NOx). In a process called "NOx scavenging," smog literally ate itself.

Interestingly enough, smog had been misnamed. The key ingredients were not "smoke plus fog equals smog," as originally presumed in London, Los Angeles, and elsewhere where the brown haze first became a nuisance, but rather "volatile organic compounds plus oxides of nitrogen equaled ground-level ozone," or "VOCs plus NOx equals O_3." These were often joined by nitric acid and sulfur compounds, which burned eyes. Smog's organic compounds were the carbon-based chemicals found in all living things and in products made from living things, such as coal and oil. Volatile organic compounds were simply the ones that vaporize at room temperature and normal atmospheric pressure. Oxides of nitrogen occurred naturally but were also a by-product of incomplete combustion, such as occurs in car engines. Both VOCs and NOx entered the atmosphere from factories, gas stations, dry cleaners, and so on, as well as from natural processes. But a big supplier of both was the internal combustion engine. In America in the 1990s, cars, trucks, and buses produced between one-third and one-half of the total VOCs and NOx that sunlight cooked into ground-level ozone and the weather shifted around. In the Los Angeles basin, where stiff regulations had actually decreased smog-making emissions from their all-time highs of the 1950s and 1960s, vehicles produced well over half the totals of both.

A Brief History of Smog

Smog was first scientifically identified by Dr. A. J. Haagen-Smit, a professor of biochemistry at the California Institute of Technology, in 1950. How Haagen-Smit reached his conclusions, how long it took for the conclusions to be accepted, and how the industries that tried to undermine his conclusions executed their nefarious tactics may all seem a little quaint after the passage of almost half a century. They're not quaint, however; they're prophetic. For although the scale of air pollution increased an unknown magnitude in the second half of the twentieth century, the means of discovery and the struggle to respond to new knowledge evolved quite slowly as a result of industrial, political, and economic forces strongly wedded to the status quo and to a strategy, called "least feasible retreat," that guarded the status quo. Least feasible retreat combined prolonged scientific fact gathering and study, legal maneuverings to prolong delays, press and media information to sow doubt and confusion, extensive political lobbying, and protracted insistence that everything—or most everything—was not as bad as naysayers made it out to be. The overall goal of the tactics was to undermine momentum for change and to cast doubt over the desirability of new paths.

Haagen-Smit was Dutch and had a European air, wore dark suits and white shirts, smoked good cigars, and referred to Cal Tech as "the institute." In the late 1940s, he was experimenting with flavor compounds, seeking ways to preserve the taste and enhance the aroma of canned pineapples. Food processors wanted to make canned pineapples more appealing. The Los Angeles Chamber of Commerce, to which farmers were complaining that their crops were being damaged—could it be the air?—approached Haagen-Smit and asked him if he would study smog. He took the job.

At the time, Los Angeles was a magnificent natural basin whose air was all too frequently suffused with a mustard brown haze that made eyes burn, throats itch, and lungs pump harder because of shortness of breath (this side effect later became known as "ozone burn"). Tourists complained that they couldn't see the mountains or Santa Catalina Island off Long Beach; they could see plumes of smoke. Visible for miles on clear days, the plumes often twisted like scarves out of tall smokestacks. Was smog smokestack industry's fault? It was an easy conclusion to draw. Plus the number of tall chimneys was increasing. Los Angeles was booming. Drawn by good jobs and year-round mild temperatures, Americans were making the city, home of movie stars and the perpetual tan, a postwar Shangri-la. In an American Shangri-la, everyone drove. The Detroit-based automakers helped. Led by General Motors, they bought the city's trolley lines, which tied the sprawling communities together like a spiderweb, and tore up the tracks.

As Los Angeles grew, so did the number of cars. Smog came more often. Of course, like many evil things, smog had a good side. It could be beautiful. It could refract sunsets into technicolor displays beyond anything Hollywood could concoct. On occasion, the sky simply glowed.

Some scientists speculated that the glow and the haze were more complex than generally acknowledged—and that cars played a role. They were ubiquitous, their emissions obvious. Yet hypotheses linking cars to smog were discounted. That was understandable. Los Angeles loved the car. The soul of this twentieth-century city was social mobility, its perfume, petroleum. The car, in all its material glory and speed, its design wantonness and ego gratification, had woven itself deeper into the psyche here than anywhere else in America, probably the world. Many new homes had a room for two cars, often under the owner's bedroom.

When the public clamored for action, officials responded by funding scientific studies. Costly and seriously debated, the studies tended toward solutions for smog elimination that leaned heavily toward technologies more appropriate for movies than for reality. For instance, one study concluded that tunnels drilled through the mountains surrounding the city would naturally carry the smog away, out over the Mojave Desert. Another study suggested using helicopters, hovering overhead, to eliminate the stuff. A third idea involved using big fans; they ought to do the trick. They didn't because the study never got funded. One reason, critics noted, was that big fans would drain Los Angeles of so much electricity that there wouldn't be enough left to run more mundane things like factories, air conditioners, and movie projectors.

Adding to the complications of figuring out why Los Angeles got so much smog was the city's peculiar atmospheric history. The first recorded atmospheric incident had occurred in October 1542. Explorer Juan Rodriguez Cabrillo, a Portuguese sailing for Spain, anchored offshore of what would one day be Long Beach and called the place La Bahia de Los Fumos, or "the bay of smokes." According to one account, Native Americans living in the basin were tending fires, and smoke was going straight up several hundred feet, hitting Los Angeles's now-famous inversion layer and sprawling like a gray blanket. On a later occasion, in 1903, an editorial in the *Los Angeles*

Herald mentioned a time when "smoke fumes . . . obscured the sun and drove out the daylight." As recently as 1941 and 1942, when an irritating haze covered the city repeatedly, rumors blamed the Japs—they were offshore in submarines, deploying gas, it was said.

Further complicating things was Los Angeles's unique topography. The city was cradled in a vast natural box, with mountains on three sides: the San Gabriels to the north, the Santa Monica Mountains to the west, the San Jacinto and Santa Ana ranges to the east. Cool currents of air blowing in off the Pacific formed the fourth side of the box. The bottom covered almost 1,600 square miles. The top was often capped with an inversion layer at between 1,000 and 3,000 feet, effectively putting a lid on the box.

In 1949, when Haagen-Smit first perched the smog-collecting devices he'd fabricated in his Pasadena lab to filter smog out of the air in some windows and turned on the fans, he was an air pollution pioneer. Very little was known about phenomena like inversions and smog, or about their synergies. Haagen-Smit did know that smog was complex, that it wasn't just smoke and fog, as had once been believed. Smoke, fog, sulfur compounds, invisible fumes, gases—they all could be part of it. Yet the basic chemistry puzzled him. None of the individual contaminants, of which there were more than fifty, had a brownish shade. Some reaction between two or more seemed the likely explanation. As Haagen-Smit got closer to solving the puzzle, the pro-industry Chamber of Commerce, fearing a conflict, withdrew from the research picture. Fortunately for Haagen-Smit, Los Angeles had America's foremost air pollution authorities working for the city. Having been given license in 1947 when Los Angeles County adopted the country's first air pollution regulations, the authorities provided the scientist with additional funding.

In late 1950, Haagen-Smit delivered his report. He had concluded that the two main ingredients of smog were volatile organic hydrocarbons and oxides of nitrogen. Cars, trucks, and buses produced voluminous quantities of both, Haagen-Smit said, but he felt optimistic that the problem could be cleaned up in "a short time." The European professor with the gentlemanly air might as well have been a prima donna wandering naively onto a football field. He got clobbered. Industry and business both went after him. A blue-collar workers' group called his findings a plot to hurt the little guy who loved his car. Worst of all, in Haagen-Smit's eyes, was the response of the Stanford Research Institute (SRI), a private research group not associated with Stanford University but paid for by the oil industry. This institute challenged the science of the professor from Cal Tech. The SRI's scientists replicated the smog-filtering experiments and concluded that "Haagen-Smit was all wet." Smog "was no more irritating than fresh outside air," they said, and, like life itself, emanated from "some mystic sort of chemical reaction."

His professional integrity besmirched, Haagen-Smit was hurt. He decided to prove the SRI wrong. He launched himself on a new career path, becoming an environmental crusader two decades before environmental worries clouded the average American mind. Taking the point position in what became California's protracted regulatory war against oil, auto, and other emission-heavy industries, he chaired several air pollution boards, including the first California Air Resources Board (CARB). As a crusader, Haagen-Smit was dubbed "Old Dutch Cleanser." As a regulator, he "was able to cajole and encourage at the same time," recalled Gladys Meade, who served under him on the Air Resources Board. One of the crusading professor's weaknesses, cigars, a product with a deadly emission profile of its own, may have contributed to his death. He died of lung cancer in the mid-1970s, a couple years after Governor Ronald Reagan, irked because Haagen-Smit refused to take orders, fired him as chairman of CARB. CARB's research laboratory in El Monte was named after him.

Back in the early 1950s, during his first bouts with industry over the makeup of smog, Haagen-Smit had pointed a finger at the oil industry and its public relations departments. The gentlemen working for the departments were perpetuating smog by clouding the relevant issues and by increasing confusion with self-serving data and disinformation, he said, with their sole purpose being to obscure big oil's crucial role in smog's formation.

It took six years for the oil industry to concede Haagen-Smit was right about smog's makeup. That happened only after an independent research group was created explicitly to resolve the ongoing imbroglio and took his side. At the time, automobiles had not been named as a big part of the problem, though they were guilty by complicity, since oil refineries made gasoline. Spokesmen for the auto companies did admit that smog was troublesome and that cars might play a small role in the problem. But they also insisted that Los Angeles was an aberration. Smog occurred only here; it was nothing to worry about anywhere else in America.

During this period of contentiousness and early scientific discovery in Los Angeles, a process to wrestle with air pollution and to promulgate regulations to mitigate it slowly and painfully emerged. The process, evolving into a pattern, would continue to be repeated right up through the 1990s. First, like Haagen-Smit, scientists would conduct research and gather substantial evidence showing that vehicles burning petroleum polluted in a certain way: at the pump, on the road, idling, and so on. Second, disagreements arose over the merits and validity of the research. Third, automakers and oil companies got more actively involved in the policy-making process to deal with the research findings. Fourth, the automakers and oil companies prolonged the debate about the research, resulting in long delays in reaching consensus and initiating action. Here was the origin of least feasible retreat, the strategy that frustrated regulators, often exhausted the finances of those calling for controls, and stalled change in the name of the status quo.

A Pestilence of Vapors Threatens Health and Climate

Ever since Haagen-Smit identified smog, scientists have debated its effects on people, plants, animals, and things, including buildings and monuments. Its effects on people have pulled in the most research dollars, which paid for a voluminous number of often complex studies with titles such as "Estimation of Risk of Glucose 6-Phosphate Dehydrogenase-Deficient Red Cells to Ozone and Nitrogen Dioxide" and "The Role of Ozone in Tracheal Cell 3 Transformation." Yet conclusive proof of the damage wrought by smog on humans remained difficult to measure with the accuracy scientists strive for and politicians need to take often unpopular stands. By the 1990s, after forty years of studies, some scientists still claimed that smog looked worse than it was. In normal people, they said, it caused only short-term effects, like burning eyes and shortness of breath during exercise. In 1994, Dr. Jane Warren, director of research at the Health Effects Institute in Cambridge, Massachusetts, a research center focused exclusively on the effects of automobile emissions, said that no direct correlations between ground-level ozone and long-term health had been conclusively proven, even at exposures beyond the national standard, which was 120 parts per billion.

"The answer is not in," insisted Dr. Warren. "There is a lot we just don't know; there is this complicated web."

Other scientists said the answer *was* in. They claimed that the integrity of the atmosphere was being compromised and that on the molecular level smog and other air pollutants changed the very structure of the lungs, decreasing lung function, making lungs rigid from prolonged exposure, and subjecting children in particular, because of their higher rate of activity, to a variety of respiratory ailments and infections. A study often alluded to by those convinced that smog was worse than publicly acknowledged had been conducted in 1990 in Los Angeles. After performing a hundred autopsies on healthy young people who lived, and died prematurely, in Los Angeles, Dr. Russell P. Sherwin, a pathologist at the University of Southern California, said, "These are pretty young people . . . running out of lung." Some had smoked, Sherwin reported, "but the damage I'm seeing is above and beyond what we've seen with smoking."

In particular, Sherwin found lesions in bronchioles, the tubes that lead to little air sacs in the lungs called alveoli. The alveoli are where oxygen and carbon dioxide are exchanged; oxygen goes in, enriching the blood, and carbon dioxide, a waste product of the body, goes out. Alveoli provide a large surface of thin cells through which the gases can be exchanged, the oxygen diffusing through the walls and entering the blood through the capillaries, the carbon dioxide diffusing from the blood into the alveoli. An average adult has about 300 million alveoli with a total surface area of seventy to eighty square meters.

In the 1990s, the assessment of health risks from air pollution was a slow, expensive, institution-monopolized specialty—a world of epidemiologists, departments of toxicology, and computer modeling. Establishing scientific proof that a specific toxin that sailed out of millions of tailpipes caused cancer or illness in a certain number of people in a predictable way was not easy work. Findings often caused tremendous anxiety, tripped off legal maneuverings, and justified large media campaigns to deny guilt. All the while, the scientific proof on which the American system relied to award payment for damages or to demand change was subject to aggressive counterclaims, just as Haagen-Smit's work had been. No industry or manufacturer wanted to have its emissions linked to what biologists called "target tissues," such as lungs, blood, and body organs. One subtext to the health issue was this: Just how much is one person's health worth in a capitalistic society? How much should the auto industry have to spend on new technologies, which all drivers would have to buy, to save a few lives?

There were a few things most parties in the emissions debates tended to agree on. One was that emissions were real and were more irritating than fresh outside air, as the Stanford Research Institute had said about smog in 1950. Second was the fact that most of the cheap and easy things that could be done technologically to lower emissions, at least in America, had been done by the 1990s; new solutions were going to carry higher price tags. Third, emissions were increasing around much of the world, especially in developing nations falling in love with automobiles and enjoying industrial growth; in fact, the number of vehicles worldwide had increased tenfold since 1950 and, together with population, continued trending upward sharply.

The Clean Air Act Amendments of 1990, which updated and broadened federal regulations, were the latest attempt to improve things. Eight hundred pages in length, compared with the original act's fifty-odd pages, the amendments added air toxics and particulates to the list of "criteria pollutants," those the Environmental Protection Agency had proven to be hazardous to health. Air toxics included dozens of chemicals, such as ammonia, dry cleaning fluid, and benzene. Particulates are micron-sized specks that float in the air. If you have ever cleaned ashes from a woodstove, sat in the smoke of a campfire, or ridden a bike in a coal-burning region, you have seen particulates clustered by the thousands when you blew your nose. Particulates, plentiful in the black exhaust of diesel-burning engines, can slip between nasal cilia, enter lungs, and become lodged in the bronchioles or alveoli or diffuse into the bloodstream.

The silver bullet of the 1990 Clean Air Act Amendments continued to be "technology forcing"—a philosophy that had emerged in the 1960s when its predecessor, "technology following," hadn't accomplished much. Simply stated, technology following was a pro-manufacturer stance, and technology forcing was an antimanufacturing one. Technology following said that manufacturers had to be able to make an antipollution device that customers could afford, and on which manufacturers could make a profit, before the device could be mandated, say, to be put on cars to lower emissions. Technology following had been part of the California Motor Vehicle Pollution Control Act of 1960, a key early bill. In the bill, because of concern that one manufacturer might gain a monopoly, the language said two manufacturers had to have affordable devices available before California could mandate them. The philosophy of technology forcing had resulted from technology following's failure to get manufacturers to act aggressively. Instead of letting manufacturers set the pace, technology forcing said that manufacturers had to meet certain standards within certain time frames. How they met them was their business. But technology forcing forced change; it didn't wait for it to happen because manufacturers responded. On the federal level, technology forcing, which originated in California in the 1960s, was validated with the Clean Air Act of 1970.

All of this was a relatively sudden change. Manufacturers were not used to having government tell them what to do, even voluntarily. Suddenly, they became scapegoats for having supplied America with the progress it wanted. It's easy to see why manufacturers got riled up. Before 1970, when the first federal Clean Air Act with teeth had been passed by an environmental-friendly Congress, air pollution had generally been viewed as a minor cold a booming America simply had to live with. Strong auto, oil, and industrial lobbies had opposed any kind of controls on the economy. Philosophically, Congress was hardly ready for controls either. The popular view was that smokestacks and mufflers wafted forth pollutants, particulates, and air toxics in a salute to American abundance, not to sicken people downwind or to eat away public monuments or to ruin farmland. Only gradually, and much against the wishes and arguments of the status quo, did air pollution take on an ominous, often difficult-to-grasp quality in the public imagination. After all, often you couldn't see it. Even when you could, say, with smog, you might not sneeze much or get short of breath unless you were exercising or had lung problems already. So, at times air pollution was easy to deny or ignore.

Not in Los Angeles, however. Both the city and the State of California wrestled with smog and air pollution long before other places did. The early battles in the 1950s and 1960s were often as much about philosophy (the government's right to intervene in commerce) and science (was it accurate enough to get a grip on the mysteries of air pollution?) as they were about taking action. In those days, taking suspected violators to court to enforce regulations was hard. Scientific yardsticks to measure damage were few and vague, and in court they were challenged by scientists and experts working for industries that were opposed to yardsticks. In addition, few technologies were available for even the most conscientious air polluters to turn to, or the most conscientious car owners to buy.

After 1970, begrudgingly at first but then moving eagerly into the new research realms that mandated emission controls created for their engineers and scientists, the auto and oil industries devised low-cost devices, eliminated lead from gasoline, and dramatically lowered smog precursors and carbon monoxide from new car tailpipes. Stationary sources of air pollution made giant strides in lowering sulfuric emissions from coal and reducing other air pollutants. Yet air pollution problems across America only began to stabilize; they didn't improve because the growth in the number of sources of emissions offset individual gains. Then, during the 1980s, automotive air pollution started to get worse again. More cars driven more miles per car, with resulting congestion, which increased emissions, were the main culprit. Meanwhile, clean air regulations stood still, and in some instances slid backward, abetted by the Reagan administration's antienvironmental stance. In the 1980s, the technology fix of choice for cleaning the air continued to be the catalytic converter, which had emerged in the 1970s as the key emission technology for cars, but with new and better catalysts. Lifestyle changes remained taboo. Getting Americans to carpool, to pay high parking fees to discourage driving, or to take public transit all had been tried in the 1970s, during the heyday of the environmental movement. They had, with rare exception, failed. Value-laden, disruptive to the status quo, and difficult to enforce, lifestyle changes became anathema to politicians, though environmentalists continued to champion them as the cheapest, most practical long-term solutions to too many cars and too much air pollution.

8

Australia, 1987: The First World Solar Challenge

Solectria IV, MIT's entry in the 1987 across-Australia World Solar Challenge, had a photovoltaic array that could be angled to catch sunlight.

James was considered a weird kid in high school. He was shunned by the popular crowd, even by the average crowd. At MIT, he had some intense friends . . . but he was a bit of a loner. You never can completely get rid of how your self-image gets built when you're young.

—Catherine Anderson, mechanic,
MIT Solar Car Racing Team

Good air, or the lack of it, might have been the main driver behind the technologies James Worden and his teammates worked on at MIT in 1986 and 1987, but getting vehicles built in a less-than-ideal environment was what made them run. Or race, as it was. For what Worden and his team really loved to do was race solar cars. "It's in my blood," Worden said.

In the summer of 1987, the team raced Solectria IV in the Swiss Tour de Sol. A Discovery Channel television crew followed every move, but the car had reliability problems. "It was quite a saga," Anderson recalled. "Motors blew up, controllers blew up."

Back in Cambridge, the team tinkered with the car. Then they flew to Australia to compete in the first World Solar Challenge, the brainstorm of Hans Tholstrup, a Danish-Australian adventurer who had flown a plane around the world solo and now was obsessed with alternative-energy technologies. Tholstrup invited teams from the United States, Japan, Denmark, Switzerland, Pakistan, and elsewhere to make solar cars, bring them to Australia, and to race them from Darwin to Adelaide. Like the cars they built, the teams were unusual. Many were led by what one journalist called "that rarest of all breeds, the successful idealist."

One such idealist was Paul Mitchell, head of the hair care company John Paul Mitchell Systems. Mitchell's Mana La, which in Hawaiian means "power of the sun," ran on a combination of wind and solar energy. Appropriately, it looked sort of like a hair dryer. Australian Dick Smith led Team Marsupial. A type of businessman you seldom saw on the other six continents, Smith had a bizarre sense of humor. Once, parodying America's Evel Knievel, Smith had jumped a double-decker bus over seventeen parked motorcycles.

From Germany came D. Schmitz, who landed in Darwin with three pieces of luggage. Inside the luggage was his disassembled solar car. He pieced it together and drove off. Schmitz was soon nicknamed "Suitcase Man."

Though less colorful, a team that united General Motors, Hughes Aircraft Company, and AeroVironment, Paul MacCready's California-based company that had a reputation for pulling off one-of-a-kind projects, brought the most advanced car, the Sunraycer, to Australia. A 390-pound vehicle built from scratch in seven months for $3 million, Sunraycer quickly became the odds-on favorite to win the race. AeroVironment had even sent a scout ahead to videotape the 1,950-mile route across the Australian outback, counting cattle grids (113), noting the gradients of hills, and recording weather patterns for the last three years. It was the kind of strategy a team could afford if GM backed it and decided it was going to win.

In comparison, the MIT team had its guerrilla engineering strategy, which had been honed by repeated fine tunings and repairs right at the starting line. It also had Solectria IV, the car whose components had blown up recently in Switzerland. The day before the World Solar Challenge, Solectria IV had a different kind of problem. It caught fire while Worden was driving. Seeing smoke billowing from the car, Catherine Anderson waved at James to stop, ran to the car, thrust both hands under his armpits, and lifted him out. Worden's hair was singed, and he was in slight shock. As flames leaped from the battery compartment, Anderson tore off the insulation. Once the fire was extinguished and James was safe, Anderson practically collapsed, exhausted, her adrenaline rush subsiding.

"It was pretty frightening," Worden said of the experience. "But it's all part of learning how to do things better." The problem had been an electrical short: the seat and batteries were too close together.

After the fire, the race itself was somewhat anticlimactic. Solectria IV broke down a lot and took ninth place, although it never finished. The Hawaiian entry, Mana La, despite having a quartz crystal repeatedly placed on it to channel some good luck down from the sky, was beset by battery problems. Team Marsupial, adventurer Dick Smith's entry, did manage to come in third, but not until four days after Sunraycer had finished. The GM/Hughes/AeroVironment entry ran the course in five and a half days, averaging 41.6 miles per hour. Cynics scoffed that it just went to show that if you had $3 million and catering you could win almost any race.

That wasn't fair, however, to the stupendous effort invested by the winners. From its inception seven months earlier, when an invitation to compete in the World Solar Challenge had arrived as a complete surprise on the desk of Howard Wilson, vice president of GM Programs for Hughes Aircraft Company, the Sunraycer team had demonstrated a sense of commitment and an ability to make a totally new vehicle that reflected very positively on GM. Blending Hughes's technology, GM's auto capabilities, and AeroVironment's innovative leadership in low-speed air dynamics with lightweight structures, the team showed what one of the Big Boys, working with an agile and small partner, could achieve. This wasn't four college kids stashing stuff in closets and getting called on the carpet for stinking up the ventilation system.

"It wasn't any wonder we made quite a good vehicle," Howard Wilson said. "We had GM's blessing at the highest level, and we had the funds."

9

1989: The Founding of Solectria Corporation

C. Michael Lewis

The name Solectria is an anagram of "sol" for sun, "elec" for electricity, and "tria" for three, since Worden's early vehicles all had three wheels.

Being a steward of a vision shifts a leader's relationship towards her or his personal vision. It ceases to be a possession, as in this is my vision, and becomes a calling. You are "its" as much as it is yours.

—Peter Senge,
The Fifth Disciple

That fall, back at MIT, Worden was test-driving Solectria V on Memorial Drive when Catherine Anderson brought a classmate, Anita Rajan, out to watch. Rajan had been born in England to Indian parents. She had gone to school in Algeria and Florida, where she'd been valedictorian of her high school class. At MIT, she was studying electrical engineering. She'd known Catherine Anderson since they had been freshmen, and through her had learned about the solar racing team. Listening to Anderson, Rajan had become more than a little curious about the brainy, hands-on leader of this dream team that traveled, took risks, and sounded very exciting. She knew how Anderson had pulled James out of the fiery car in Australia, how she'd occasionally tucked him in before races to make sure he got enough sleep (he'd still get up, come over, and poke around). She knew Anderson was doing interesting, real-world stuff, like machining parts and working with batteries, for the racing team. And though Rajan knew little about these things herself or, for that matter, how to sneak into MIT's labs at night to mill parts, she memorized a few solar racing buzzwords so when Anderson introduced her to Worden she had a handle on some of the lingo.

Worden must have been impressed. Soon he and the dark-haired, reserved Anita Rajan began dating.

"I think they knew they were going to get married the second day they met," Anderson said. "Both were very driven, focused people. They had similar backgrounds and values. When they believed in something, they believed in it very passionately."

Worden took Rajan into the labs, taught her how solar cars worked, explained batteries. She became project manager for the team not long after the American Solar Cup, a race held in Visalia, California, in September.

In the fall of 1988, solar and electric car racing in America was in its infancy. Few people had ever heard of these cars, much less seen one. There were no zero-emission mandates to hurry along attempts to clean America's air (they would come in 1990); there were no Pentagon "dual-use" dollars for developing EV technologies for both the road and the battlefield (that would happen after the Persian Gulf War in 1991); there were no *Business Week* covers with EVs on them (that would be in May of 1994). But there was a core of enthusiasts and tinkerers committed to clean-car concepts, and a common inspiration for them was AeroVironment's CEO, Paul B. MacCready Jr.

AeroVironment was located in Monrovia, California. Its motto was "Doing more with less." The company's literature claimed that engineers there imagined the impossible and did it. AeroVironment projects included design of the General Motors Sunraycer; design and construction of the Solar Challenger, which flew from Paris to

England powered only by the sun; and creation of a replica of *Quetzalcoatlus northropi,* a giant pterodactyl, for the IMAX film *On the Wing. Quetzalcoatlus northropi,* or "QN, the Time Traveler," had three sensors and thirteen electric motor muscles linked to an electronic brain so it could fly right.

In September 1988, when James Worden was showing Anita Rajan how solar cars worked and C. Michael Lewis, the guiding spirit behind the Solar Car Challenge, was finalizing his event, in AeroVironment's labs the hush-hush design of the Impact, an EV born of the Sunraycer and destined to be launched as GM's EV1 in 1996, was being refined. So when Paul MacCready took time to get away and travel to the American Solar Cup in Visalia, he lent the event a certain cachet.

Still, there were only six entries, and just two completed the 150-mile circuit. One was an improved version of Solectria V, a DOW Chemical Company decal stuck on its bullet-shaped nose. Behind the wheel, across a flat, hot stretch of road, Worden nursed the car up to ninety miles per hour—a record for a solar car. "That car didn't handle perfectly," Worden said later, "but as long as you kept a handle on it, it was okay."

California state troopers were posted along the route because it was dangerous running so fast through towns. Worden flashed by them, going extremely fast in what looked like a silver bullet. Had they been in the audience, MIT administrators would probably have turned white. A photograph taken after the race shows Worden grinning from ear to ear, pumped, getting a speeding ticket from one of the troopers.

In those heady days, working on solar-electric racing cars was an all-consuming passion for both James Worden and Anita Rajan, who took to the vehicles' promise and magic almost as strongly as he had. Movies, museums, goofing around—such youthful diversions were out; they required too much time.

"Sometimes, we pulled all-nighters with a race coming up," Anita Rajan said. "I think in the long run all of us learned a hell of a lot doing it that way. Just the same, I wish we had spent more time just goofing around, instead of working so hard on cars that, in many ways, the school got more recognition for than we did."

The shift from an academic, clublike environment to a start-up came in 1989. In May of that year, just about the time of Worden's graduation, the first American Tour de Sol took place. Modeled after its Swiss predecessor, the race went from Montpelier, Vermont, where Governor Madeleine Kunin waved the vehicles off with an Earth flag in front of her state capitol building, to Cambridge, Massachusetts. MIT's entry, a modified Solectria V, won the event with Worden driving.

Then solar and electric car developments in America got a big boost from GM. Still pleased with Sunraycer's victory in the World Solar Challenge, which had generated tremendous publicity for the often maligned corporation, and deep into creating the first Impact prototypes with AeroVironment, GM announced that it was sponsoring a new event called the Sunrayce. It would go from Orlando to Detroit. GM offered seed money for college engineering departments to design and enter vehicles; all they had to do was submit proposals and get them approved. The concept immediately caught the attention of universities looking for engineering challenges with some free start-up money and seeded a growing cadre of solar and EV devotees.

In Massachusetts, James Worden and Anita Rajan started to make plans to enter a special car Worden was building, an odd-looking, ravioli-shaped thing called the Galaxy. Their goal was to set a land speed record on the Bonneville Salt Flats in Utah. The Galaxy never made the Sunrayce, though. Worden and Rajan decided that more racing for an MIT graduate who needed a job and for his girlfriend, now a senior, wasn't smart. What was smart was becoming a supplier to the universities and colleges eager to enter the Sunrayce and other races. "We scraped together a catalog, got it xeroxed, mailed it out to all these universities," Rajan said. In October 1989, they made their first sale, a solar panel.

Suppliers warmed to the concept of two young engineers with credentials offering to be middlemen between them and universities and students, who were notoriously difficult to deal with because they asked endless questions and never had money. Getting in the middle "was a great way for us to market ourselves," Rajan said. After the first sale, business "grew slowly, but kept going."

10

Forces Gathering Behind Cleaner Cars and Air

Arkie G. Hudkins, Jr. *Automotive News*

Automotive News often criticized the California zero-emission vehicle mandate. This cartoon, with its clichéd view of Californians, ran twice.

> **In the late 1980s, every day a new story came on the scene to heighten public concern, from Chernobyl to toxins to ozone depletion to the greenhouse effect. They kept things rolling.**
>
> **—Jananne Sharpless, chairperson, California Air Resources Board, 1985–1993**

Elsewhere—in California, in Washington, in the Northeast—forces behind cleaner, greener vehicles were intensifying.

California's Zero-Emission Vehicle Mandate

In California, America's laboratory of atmospheric sickness and health, pressure was once again building to do something about declining air quality. During the mid-1980s the state's economy had boomed. Many new jobs were created in clean, light industries and in government services, a highly desirable mix. One negative consequence, though, was an economic growth/clean air paradox: the creation of clean jobs was offset by the pollution from vehicles getting people to the jobs. Ultimately, more vehicles driven more miles, often by commuters living farther and farther from their jobs, overwhelmed the gains that had been won with emission-control devices added to individual vehicles.

Neither environmentalists nor public health advocates let Californians forget how all their driving hurt everyone's health and, as scientists were beginning to conclude, might be changing the world's weather as well. In addition, economic sanctions loomed if the state continued to fail to meet federal clean air goals.

California was in an old place: either meet federal clean air deadlines, which were beyond any reasonable chance of being met, or invent a new path toward a hypothetically cleaner horizon. The conundrum was nothing new. For almost forty years, California had been setting, and trying to enforce, the most progressive air pollution regulations in America. But whether the regulations had followed technology, as they had early on, or forced technology, which they had after 1970, they never seemed to catch up to the increasing number of cars.

It was frustrating, recalled Jananne Sharpless, chairwoman of the California Air Resources Board between 1985 and 1993. "In the late 1980s," she said, "Californians had to admit again that we weren't going to meet the federal deadlines. We had to put together state implementation plans, called SIPs, that recognized our shortcomings. Then we tried to negotiate 'reasonable further progress' with the federal Environmental Protection Agency to meet the 1989 goals."

"Reasonable further progress" was a euphemism. It meant, "No way in the world can we do that, but we'll give it a shot—just don't penalize us!"

Reasonable further progress flopped. Other states were also having trouble complying, but they adopted a simpler solution. "They lied," Sharpless said. "They put together what we called 'cheater SIPs.' They couldn't meet the deadlines either, but said they could."

Enforcement of clean air goals by the EPA during the lost decade of the 1980s was lax. Many states got away with cheater SIPs. California, however, couldn't. The leader in America's clean air battles on both automotive and industrial fronts since Haagen-Smit's days, it couldn't cheat because too many environmental groups kept an eye on things. Still, "regardless of everything we were trying to do, we were sort of being overwhelmed," Sharpless said. "What were we going to do? For the last twenty years we had continued to ratchet down and ratchet down on the internal combustion engine to greater and greater stringency at greater and greater cost. If we didn't go after mobile sources [mainly cars, light trucks, commercial trucks, and buses], and go after them in a big way, industry was going to have to pay a higher and higher price. And we'd already gone after big industry. The message became, 'Everybody has to pay.'"

But nobody volunteered.

Desperate for a way out of having to pay or change its mobile lifestyle, California looked around for alternatives. One with promise was methanol.

For a while, in 1988 and 1989, it looked like methanol was the alternative fuel that was going to save California's air. It was going to help farmers because it could be made from grain. It was going to diversify the state's energy diet, which was petroleum rich. Introduced when environmentalism was politically hot and, as expected, opposed by the oil companies, methanol's promise prompted the creation of a study panel chaired by Sharpless. Members came from auto and oil companies, from farm groups, from the American Lung Association. Charged with finding a "fuel-neutral" direction for California, the panel used methanol as a benchmark.

Methanol emitted less carbon monoxide when burned. Carbon monoxide, one of the EPA's criteria pollutants, impaired the blood's ability to carry oxygen to cells and tissues. At high concentrations, like those found in people suffocated by carbon monoxide, it made the brain reek of an ammonia-like smell and made organs smell like garlic. Methanol also emitted fewer hydrocarbons than gasoline, and its emissions were less reactive in the atmosphere. In California, that meant methanol could lower the formation of carbon dioxide and dampen the greenhouse effect. But methanol also had shortcomings. Its processing created more carbon dioxide than was saved by burning it as a fuel. It had a low energy density, approximately half that of gasoline; it took two gallons of methanol to drive the same distance as one gallon of gasoline. Methanol also had a high corrosion factor (it ate away standard fuel tanks and rubber gaskets) and high toxicity (swallowing the fuel could kill you).

While the profile of methanol was emerging under rigorous study, a new type of conventional gasoline, known as "reformulated gas," was conceived. Sharpless, in the thick of the alternative fuel debates in California, said that the oil companies both supported and disliked the concept of reformulated gas. They didn't like the fact that refining it would require them to invest tens of millions of dollars. They did like it better than methanol and other alternative fuels, which they did not control.

"The situation had the oil companies really kind of spinning around," Sharpless remembered. "Ultimately, they saw the handwriting on the wall. They saw the environmental momentum building behind alternative fuels. If they didn't get on board behind reformulated gasoline, they might get stuck with something they didn't like. Their thing was, 'If we can keep the State of California fuel neutral on this issue, and then work with the car companies in some way to come up with a clean gasoline, we can kill methanol.'"

And where were EVs as a means of cleaning the air at the time? "Electric vehicles were not really high on radar screens," Sharpless said. "In 1989 the oil companies were more concerned about what was going on with alternative fuels. Together with the auto companies, I think they never really believed that the EV thing would ever become a significant happening."

Three converging factors helped make EVs significant in California. First, the California Energy Commission, looking for ways to reduce dependency on imported oil, said EVs were worth investigating. Second, politicians, in a state suffering deep defense cuts as the cold war abated, embraced EVs as a way to create new jobs. And third, the environmentalists and public health advocates kept up the chant that EVs meant cleaner air. The result was the California zero-emission vehicle (ZEV) mandate, with its 2 percent rule for vehicles sold in California by the Big 7 in 1998.

"I called it 'the jolt,'" AeroVironment's Paul MacCready said about the mandate.

The jolt certainly caught the auto industry's attention. California was the largest car market in the world, and it set trends. California was the pacesetter for what was coming down the pike in the twenty-first century. Now California said EVs were going to be a big thing, and that the Big Boys had to start doing something positive about their oil dependency.

The Clean Air Act Amendments of 1990

While the ZEV mandate and reformulated gasoline were being thrashed out in California, in Washington Congress was wrestling with the latest revision of the Clean Air Act. From its inception in 1963 under the liberal Kennedy administration, the act had been a continuous focus for strident debate about the role of government regulation in a capitalistic society. The original act had been as much a reaction to unexpected and dramatic events as a foresighted attempt to get a grip on a growing problem. Three widely publicized events in 1962—the publication of Rachel Carson's *Silent Spring,* an atmospheric alert in the coal- and steel-making city of Birmingham, Alabama, and a killer fog in London that took the lives of 300 people—had given the Congress reason to act, or react, as it was. Congress ultimately passed a toothless bill, however, one lacking any enforcement provisions. That changed with the amendments of 1970, which endorsed technology forcing, set clean air standards, and defined sanctions. A new federal agency, the Environmental Protection Agency, was created that same year, and one of its jobs was to enforce the tougher act. The act was further amended in the mid-1970s, then bitterly fought over during the early 1980s. By the late 1980s, dozens of cities across America were found to be in violation of a variety of the standards, especially the one for ground-level ozone. The Reagan administration, supported by industry, effectively stalled attempts at revision. Then, in June 1989, with Yellowstone Park in the background, recently elected President George Bush announced that his administration was going to do something about America's air. Given that Bush had been Reagan's vice president for eight years, this was a big surprise, as well as a complete political reversal. Seventeen months later, after what California congressman Henry A. Waxman called "one of the longest—and hardest fought—legislative battles in recent congressional history," President Bush signed the 800-page regulatory monster into law.

The revised act redefined the roles of the states and the federal government, named new goals and ways to reach them, gave the EPA new enforcement muscle, and introduced new and more flexible market-driven approaches to get results, including economic incentives to clean up the air and credits if a company did more than it was required to do. The battle against smog, which had gotten dramatically worse in America's air sheds and around the globe, remained a high priority. Smog would be attacked by three tactics: cleaner fuels, cleaner cars, and an inspection and maintenance system that would keep cleaner cars cleaner longer.

All in all, the 1990 revision, like its predecessors, had grand goals, more definitions, new incentives. But it also clung to the same old philosophy. It said technology would clean up America's air. It avoided the tough issues: making it expensive to drive, getting Americans out of their cars, and challenging the car's dominance of a mobile future.

Smog in the Northeast Air Shed

One region that supported the amended act was the Northeast. Politicians got behind it, primarily because of smog, which was no longer just "California's problem."

The summer of 1988 had been especially bad. In Boston, for instance, Dan Greenbaum, commissioner of the Department of Environmental Protection, had found himself repeatedly going on radio and television. On thirty days, he made announcements that smog levels in the Boston area exceeded federal standards. Bronchiole-sensitive types better stay in, kids with asthma watch television, joggers forget it.

The same year an air quality organization, the Northeast States for Coordinated Air Use Management (acronym: NESCAUM; membership: New England states plus New York and New Jersey) analyzed the federal air control plan for vehicles and was not impressed. "We found there were many, many shortcomings," said NESCAUM's director, Michael Bradley. Member states decided there were big clean air gains to be had by copying the plan California had in place. This was the pre-1990 plan, before the ZEV mandate.

All this while, as the smog thickened and new ways of cleaning the air in California and across the country were being debated and defined, progress was being made on EVs and their components. The progress culminated, in a sense, in early January 1990 when General Motors's CEO Roger Smith unveiled an Impact prototype EV at the Los Angeles International Auto Show. The Impact, destined to become the EV1, blew away the public. It even impressed the automotive press.

The two-seater went from zero to sixty miles per hour in 8.5 seconds, had a range of seventy-five miles on a full charge of its lead-acid batteries, and could be recharged in two to three hours. According to market research conducted by GM, the Impact would meet the needs of about two-thirds of America's drivers on a typical day, since they drove fewer than fifty miles. At night they could plug it in at home.

The appearance of an electric supercar almost out of the blue seemed too good to be true to EV supporters. And it was. Nevertheless, the Impact would continue to command center stage in EV circles for years as GM refined its technologies. Certainly, the Impact's introduction in Los Angeles, the world's capital of smog, in 1990, during ongoing negotiations over the Clean Air Act Amendments in Washington, suggested that GM intended to be a leader in the manufacture of attractive, competitive electric cars, leaving the competition in its emission-free dust.

"When I first heard about the Impact I thought, I'm out of electric cars," recalled Norm Salmon, a young engineer who would later work for Solectria. "I thought, they've got it wrapped, they've done it—end of game!"

To Jananne Sharpless and the California Air Resources Board, the Impact was a good sign. The car showed the board that EV technologies were out there and could be united in a fine car; all it took was commitment.

Inside GM, though, the commitment was shaky. There was tremendous trepidation about the Impact, its value, its message. After a press conference extolling the virtues of the futuristic car, CEO Roger Smith hinted at possible problems down the road. To Bill Sessa, a spokesman for the CARB, Smith said, "You guys aren't going to make us build that car—are you?"

The answer, delivered nine months later with the ZEV mandate, was yes, we're going to try.

11

1989–1992: Solectria's First Years in Business

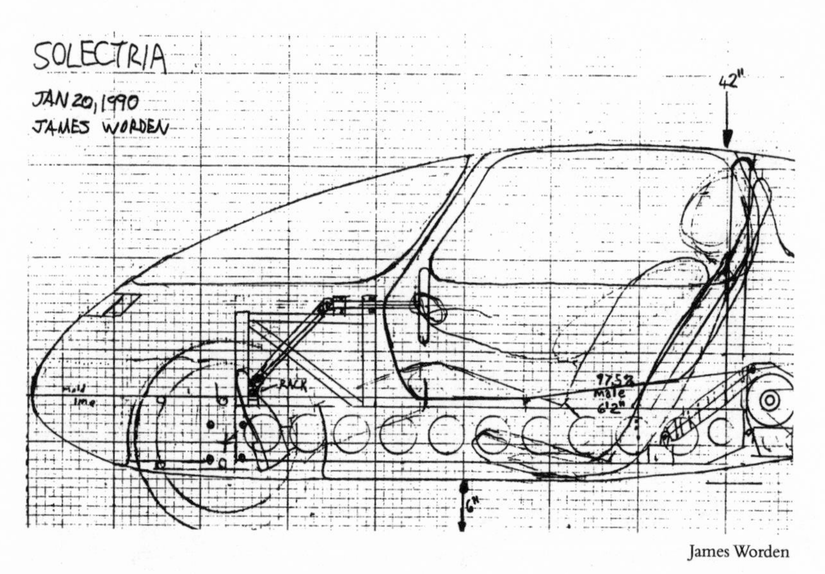

James Worden

Worden's sketch, done in 1990, of the front end and driving compartment of the Lightspeed, a four-wheel car with an aluminum frame and composite body.

If it was up to James and me, Solectria would have gone straight down the toilet.

—Ed Trembly, Solectria jack-of-all-trades

The synergistic forces—California's ZEV mandate, the Clean Air Act Amendments of 1990, the Northeast states' seeking a clean air model and finding it in California, and GM implying it had a great clean car in the works for the future—all helped tiny Solectria Corporation keep afloat in a business atmosphere of subdued optimism. The opportunities for success, both James Worden and Anita Rajan realized, now went well beyond supplying inquisitive university teams with parts and advice. Extrapolating from the ZEV mandate and from the momentum that seemed to be building behind EV technologies, it seemed reasonable to speculate that tens of thousands of EVs might be on America's roads at the end of the century.

By late 1990, after only a year in business, Solectria had its sights on such a possibility. The small company had established its strength as a technology innovator. It had proven especially good at making DC-AC converters and motor controllers. DC-AC converters changed direct current that flowed from the batteries to alternating current, which ran the motor. The more sophisticated EVs ran on AC drive. The motor controller was the "brain" of an EV. Somewhat analogous to a carburetor in a conventional car, but more sophisticated, it controlled the flow of electricity between the batteries, motor, and other components.

In 1990, Solectria also built its first company car, the Lightspeed. Designed by James Worden and sponsored in part by John Paul Mitchell Systems, the hair products company that had financed Mana La in the Australian World Solar Challenge three years earlier, the Lightspeed was a 1,200-pound sports car with gull-wing doors, a range of 120 miles on lead-acid batteries, and impressive acceleration: zero to sixty in just over eight seconds. In other words, performance-wise, it was as good as or even better than GM's Impact prototype. As a real car, though, it was only half baked. The Lightspeed lacked amenities, like a heater, air-conditioning, and cruise control. You couldn't just jump in and drive off. First you needed a tutorial on switches and gauges. So despite the fact that this bare-bones prototype won the commuter class in the second American Tour de Sol in May 1990, it was not a practical vehicle the small company could start duplicating and sell to the scattered fringe EV consumer market.

The search for a more practical vehicle to sell was complicated by the fact that initially Worden wasn't interested in conversions. A conversion was a compromise; you took a gasoline car, pulled out its engine and some systems, and put in an electric motor and other systems. "Those days James was absolutely convinced conversions were a waste of time," said Ed Trembly. "We had to build a ground-up car."

But the pressure was on. Solectria needed cash. It was surviving as much on youthful energy as it was on balance sheets and cash flow. Much of the energy came from MIT engineering students. Like filings to a magnet, they were attracted to Solectria's blend of technology forcing, high ideals, and nitty-gritty hands-on work that they could brag about back on campus. The company had outgrown the Worden family's backyard in Arlington and moved to a shop in Waltham where workdays were long, sometimes stretching to 3 A.M. Gill Pratt, former MIT racing team member and now an MIT professor, was a paid consultant. Ed Trembly had joined Solectria as its first full-time employee. The self-deprecating Trembly, who had befriended Worden years earlier when he'd shown up at an Arlington machine shop with his goofy-looking aluminum thing and a lazy Susan, told those who asked about his job, "I sweep the floors." Ten years older than James and Anita, Trembly filled the role of the wise old man. He gave the whirlwind operation, which at times threatened to collapse, some maturity, wisdom, and mechanical know-how. "It could have flopped any time," Trembly said. "If Anita hadn't been around, the company would have fallen flat on its face."

But Solectria needed more than eager student help, an electronics consultant, Rajan's finesse at paying the bills, and Trembly's know-how if it was to grow. But where did you begin if you wanted to make cars, if you wanted to be a player in the largest industry in the world? Did you build prototypes and try to get bigger companies interested in manufacturing them? Did you do conversions? Or did you simply read the history books, noting that the last company to enter the car-making business in recent times was Honda, and that Honda had made the move from a contiguous base of operations, making motorcycles, and gracefully withdraw?

James Worden didn't want to be a supplier, though he was surviving by selling components. He didn't want to be a consultant. And he didn't want to do conversions. He ended up doing conversions, however, because they made sense.

Environmental investor Jay Harris helped. In 1989, Worden and Rajan had met Harris at the Solar and Electric Vehicle Symposium at the Wentworth Institute in Boston. Completely unexpectedly, Harris posed the question every entrepreneur loves to be asked, "Would you like some money to do what you want to do?"

"It was just exactly like that," Anita Rajan recalled. "I remember thinking, 'This guy is crazy.'"

A so-called investment angel, Harris liked clean technologies, liked Worden and Rajan, and hated Detroit. He became the sole outside investor in tightly held Solectria.

Available money alone, though, didn't convince Worden it should be spent on a conversion project. There were other influences as well. Ed Trembly wanted to try a conversion. He told James, "Why don't you lower your sights a little bit? Don't be such a dreamer." Ken Sghia-Hughes, a part-timer and a fellow MIT graduate, applied constant and gentle pressure on Worden.

"James wanted to be convinced, basically," said Sghia-Hughes. "I was the one there to convince him he ought to do conversions first, as a stepping-stone. If it hadn't been me, it would have been somebody else. I kept telling him we needed to get conversions going first, work out all the electrical bugs, get a market going, get some cash flow coming in—then we could worry about doing our own vehicle."

To win their case, one weekend Trembly and Sghia-Hughes stripped the engine out of Sghia-Hughes's Geo Metro. They wanted to see how electric components would fit. They fit fine. Much to Sghia-Hughes's chagrin, the gas Geo never ran quite as well afterward, but the demonstration worked. Worden agreed to try a conversion.

In March 1991 a Geo Metro bought with Jay Harris's money was converted into the first Solectria Force. It was the beginning of an alternative-vehicle car line. At the time other small companies were converting cars and trucks to EVs, so the Force was nothing new. What was different about it was the quality of the finished car. It was a "fully created vehicle," Worden said, not just a Geo Metro with Solectria components under the hood.

The Force did have a problem, though. It howled. "That car howled something fierce," Trembly said. A lightweight transmission Worden had bought off the shelf was at fault. Trembly immediately began working on it.

Meanwhile, Solectria entered the Force in the first annual Arizona Public Service Electric 500 in Phoenix, a showcase for clean-car technologies. Noisy or not, the vehicle performed well. A conversion car with its own logo and an accessible, clean, easy-to-service motor and controller assembly, the small sedan attracted the attention of some California Air Resources Board representatives at the track. They were in Phoenix to check out how much progress was being made on the ZEVs they had mandated for 1998. Impressed with the Force, they invited Worden to bring the car to California for testing later that year.

Before the trip West, the car raced in the 1991 American Tour de Sol and was damaged when Anita Rajan, who was driving, got cut off by the MIT solar car. Rajan slammed on the brakes, and a part of the Force's noisy, lightweight transmission bent. Ed Trembly and the crew fixed it overnight, and the Force went on to win the commuter class, but after the race the transmission was replaced by a new Trembly design.

When the Force arrived at the Haagen-Smit Laboratory in El Monte that fall, it ran quietly. On the dynamometer used to measure emissions, it performed better than any vehicle the CARB had tested to date. Promptly, the Arizona Public Service Board, the Sacramento Municipal Utility District, and Southern California Edison, all of which needed to start adding clean vehicles to fleets to comply with federal law, ordered nine Forces from the Massachusetts company with three full-time employees.

Six months later, still relying on a skeleton crew, Solectria finished the last of the cars in one intense two-week rush. It was the old pattern again, a carryover from the guerrilla engineering days. Worden, Trembly, Rajan, and others worked literally around the clock day after day. "Looking back, I don't know how we did it," Anita Rajan said. "It wasn't like we were just building vehicles. We were talking to people, answering the phone. We had about fifty calls from the media the first two weeks of December. James doesn't like to spend time with the press. He thinks about them in numbers."

Line 'em up! We'll talk to 'em all at once! was Worden's philosophy toward the press.

"I was ready to scream," Rajan said.

Newcomers helped keep the lid on. Wayne Kirk, a former bicycle racer with a master's degree in aerospace engineering from Boston University and some knowledge of composites, joined the company to get the work done. His wife, Tisha, took over some of the office responsibilities. Scott Hankinson, a friend of Ed Trembly, also pitched in. Twenty-one, with visions of himself as a Wall Street stockbroker, Hankinson thought, "I'm not looking for a real job yet; I'm going to work for Solectria for a while." Hankinson went on to become vehicle production manager. Kirk would become chief engineer of the Sunrise project.

Once the Forces were all delivered, "We had a month to recoup and figure out a new strategy," Rajan said.

They needed it. In 1992, public utilities began ordering electric cars in larger numbers for alternative fuel demonstration projects mandated by the Clean Air Act Amendments of 1990. Solectria and other small conversion companies stepped in to supply the vehicles.

The Big 3 were reluctant to enter this market until sales provided the larger volumes they needed to generate profits. One of the small companies was U.S. Electricar in Santa Rosa, California, which reorganized with a management team headed by Ted Morgan. Prior to joining Electricar, Morgan had spent sixteen years with Xerox Corporation and had worked for several rapidly growing superstores. Under Morgan's direction, and with an emphasis on acquisitions and strategic alliances that courted fleet buyers, Electricar began to grow, from only 35 employees in 1992 to nearly 300 in 1994. In Florida, Bob Beaumont, whose Sebring Vanguard Company had been the world's largest street-legal electric car manufacturer back in the 1970s, was trying to get Renaissance Cars off the ground. The Renaissance Tropica, a two-seat electric car like GM's Impact, would soon be in dealers' showrooms with a sticker price of under $15,000, Beaumont claimed.

Ford and Chrysler were not totally avoiding electric cars, but their projects were relatively small and out of the spotlight. Together with GM, they had founded the United States Council for Automotive Research. USCAR's motto was "Sharing Technology for a Stronger America." The alliance's purpose was to get the Big 3 to cooperate on precompetitive research. Throughout 1992, the green light stayed on for the production of the Impact, too. A manufacturing location was even announced. Behind the scenes, though, the Impact was losing support, both within GM and in Detroit, where a backlash was gathering against zero-emission mandates, which had spread from California to Massachusetts and New York.

Nevertheless, throughout 1992 it looked like the emerging EV industry was on a roll. At Solectria "it was an amazing year," recalled Anita Rajan. "Everything came into place."

There was one scary mishap, however; James Worden almost died in an accident.

Phoenix, 1992: James Almost Dies in Battery Accident

Oh, damn, Ed Trembly thought. No, it can't be. But nothing else smells like bromine gas.

For days, while Trembly and other Solectria crew members had worked on the race car, engineers for Johnson Controls, which supplied James Worden's Force with advanced zinc-bromine batteries, had been trying to get a computer system that monitored the batteries' pumps to work right. Worden raced anyway, controlling the pumps manually from the driver's seat.

Behind him, one of the battery supply lines, which resembled a huge pile of plastic spaghetti, popped off. If the computer system had been working, it would have detected the leak and shut everything down. Instead, liquid bromine spewed around and leaked down on the hot track. Worden, driving seventy-five miles an hour, didn't notice.

By itself, bromine liquid is pretty harmless. But it vaporizes at 125 degrees, and then it can be lethal. That day the temperature on the track was over 125 degrees. Driving a second Solectria vehicle behind Worden, Trembly saw the red light come on. It meant, stop where you are! Trembly stopped, sniffed the air, and got frightened when he realized what he was smelling.

"This shaggy guy with a video camera comes up and sticks the camera in my face and says, 'So, what happened to the zinc-bromine car?' All I could do was give him some blather about it being a pretty safe car. I really didn't realize how serious things were until Anita came running down the track in tears."

Gill Pratt's wife, Janey, a medical student with the Solectria team, saved James's life. He had been overcome by bromine fumes. He just managed to pull over before passing out. The paramedics on the scene, who didn't know how to treat him, wanted to airlift him to a hospital. Janey Pratt convinced them to give Worden bronchial dilation drugs, so he could breathe.

Meanwhile, a big gas cloud was forming from the spilled bromine vaporizing off the hot track. Someone started yelling, "Run for your life!" Spectators in the stands panicked, fleeing in all directions.

"It was scary," Arvind Rajan recalled later. "James couldn't see or hear. He was in intensive care."

For Arvind Rajan, Anita's younger brother, the incident was a less-than-reassuring introduction to Solectria in the field. A graduate of Stanford University with a degree in economics and a couple years of business experience under his belt, Arvind Rajan had come to Phoenix to talk to Worden about becoming the company's first real businessperson. Instead, as a helicopter flew in to pick up Worden, Arvind Rajan watched emergency crews in white suits stalking around the hot track with boxes of Arm and Hammer baking soda. Refusing to listen to the people from Johnson Controls, who had the ways and means to neutralize the complex electrolyte, the emergency crew spread baking soda on puddles. It accomplished nothing.

Later, at the hotel where the Solectria crew was staying, Ed Trembly joined a few dispirited engineers from Johnson Controls in the corner of the bar. A minor engineering problem had become a public relations disaster. At least Worden's air passages were clearing up at the hospital, and he was going to be okay. As Trembly and the engineers talked over what went wrong, the crew from another team came into the bar and began berating the guys from Johnson Controls. They said the zinc-bromine batteries were dangerous, that the Johnson Controls engineers had behaved irresponsibly. A fight almost broke out.

"The accident never should have happened," asserted Anita Rajan. She claimed that the Johnson Controls engineers lacked racing experience and were toying with technologies without adequate safety precautions.

Back in the shop, Solectria's few other employees and the part-timers eyed Worden when he returned. Were there any aftereffects? Luckily for Worden, breathing a potentially deadly dose of bromine fumes didn't slow him down—or at least didn't have any discernible effects.

Putting a positive spin on things, Trembly said, "Actually, the whole experience was good for James. He had this little cough. Some doctor had told him it was from breathing acid fumes. After the zinc-bromine exposure his lungs were the very picture of health. Zinc-bromine gas cleaned them right out! He was in good shape after that."

Maybe. Worden himself declined to talk about the incident or the effects. His slight cough did return, however.

The accident in Arizona highlighted the danger of experimental batteries, a number of which were in various stages of development around the world. Many of them made the conventional lead-acid battery, which had environmental shortcomings of its own, seem rather benign.

12

A Regulatory Minefield: The Ozone Transport Tale

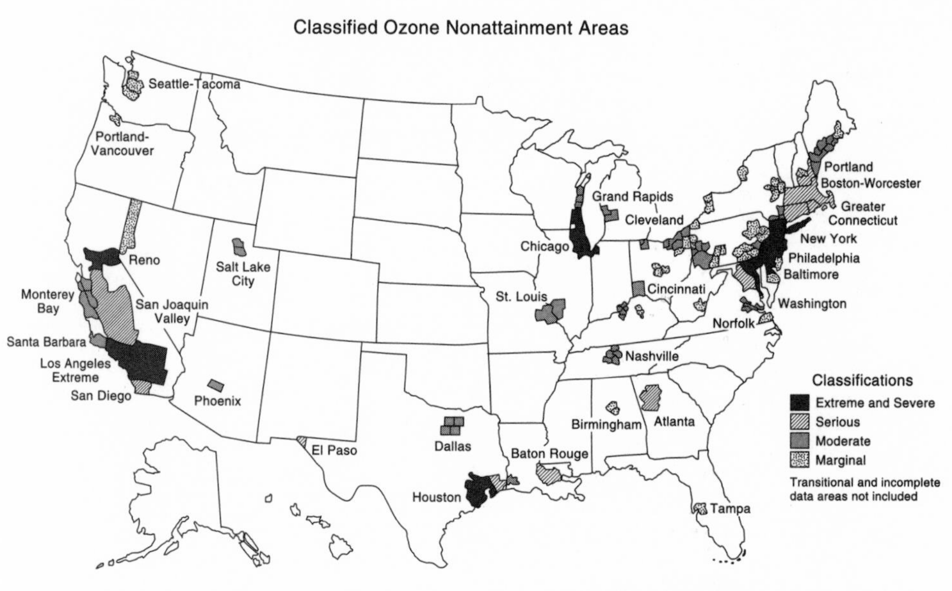

U.S. Environmental Protection Agency

In 1994, more than ninety American cities were in violation of the federal ground-level ozone standard. Many of the cities were concentrated along the eastern seaboard.

In this melodrama, we all wear gray hats. It is our own values, our desire for economic success, our demands for food, for clothing, for shelter, and for luxuries, and our desire to leave our children a little better off—all these values create environmental problems.

—Richard Brooks, *Green Justice*

Throughout 1992 and 1993, Solectria and the emerging industry in which the company was a youth-driven player with a technological edge and a lot to learn about business came into being not only because of air pollution and its impact on technology development but also because of the Pentagon's need for quieter machines and America's need for new jobs. The Pentagon's need sprang from the Persian Gulf War. After the brief war, a light came on in the heads of some Pentagon planners. Electric vehicles would make great reconnaissance machines in the desert, especially at night: they left no heat signatures out tailpipes, they were quiet, and the enemy would have a hard time detecting them. Subsequently, the Pentagon's Defense Advanced Research Projects Agency (DARPA) began investing in EV technologies with so-called dual use. "Dual use" was a new buzzword, which meant the resulting technologies had to be good for both war and peace, thus justifying the budget request from Congress.

While the Big 3 were ambivalent about EVs, DARPA marched right in under the dual-use banner and invested $68 million between

1993 and 1996. By defense spending standards, the amount was chicken feed. For small, hungry start-ups like Solectria, where most of the money went, it proved to be a veritable feast.

Also fueling the Pentagon's interest in EV technologies was the issue of energy security. A variety of disturbances, such as another emergency in oil-rich Kuwait, or a natural catastrophe like an earthquake in a major oil-producing region could disrupt America's oil flow. A disruption meant economic chaos. Consequently, energy security had been a national priority since the 1970s, when a series of oil crises had brought America to her knees. During that time, the price of gas at the pump had roughly quadrupled, and politicians had vowed never again to allow imports to exceed 10 percent of national needs. Conservation efforts between 1977 and 1985, and a doubling of the average mileage per gallon for new cars between 1973 and 1986, due mostly to federal efficiency regulations, had cut imports from the Persian Gulf by 90 percent. But the resolve to keep the imports low had gradually expired. By the late 1980s, oil imports had soared to near-record highs, helping usher in the Persian Gulf War.

The prospect of new jobs, not energy security worries or air pollution, was what turned politicians on about EVs. California, New York, and Massachusetts were all states where Americans made lots of bullets and bombs, and conducted war research. As the cold war ended and defense spending declined, these states had lost jobs. Politicians who voted for zero-emission mandates in the three states were not just thinking about clean air. An EV industry would bring jobs, high-tech jobs for which pools of labor already existed. In the big picture, though, such job creation was a win-lose situation. For instance, if California began making and selling electric cars, some other state's factories making conventional ones would lose. In all likelihood, the new would displace the old rather than augment it. This threatening trade-off fed the Big 3's growing dislike of an alternative-vehicle industry. The Big Boys didn't want green cars stealing an already shrinking number of jobs in the Rust Belt.

At Solectria, technology was the prism through which all the forces behind EV developments merged into one high-intensity beam. The challenge at the young firm, as James Worden recovered from his battery accident and orders for conversion vehicles kept coming in, was to keep the beam aimed and financed as the organization grew and learned.

To help Solectria's growth as a business, Arvind Rajan joined the company in mid-1992. Having witnessed his future brother-in-law's near death in Phoenix, Arvind Rajan may have felt the company really needed him. He certainly saw that the company needed his business acumen. Named vice president of business development, he observed that the small number of employees were doing many jobs that should be subcontracted out. He identified divisions of labor. He brought in computers. He brought in a suit. He hung it up, wearing shorts most of the time at work, just like everyone else, but at least he had a suit at work that he could pull on when necessary. He also improved the hiring process, the work flow, and the financial records. In his spare time, Rajan tried to keep pace with government affairs.

That was impossible. In the wake of the Clean Air Act Amendments of 1990, which mandated clean fuel cars in smoggy cities, and the Energy Policy Act of 1992, which forced government and utility fleets to start buying cleaner vehicles, the regulatory complexities were becoming increasingly difficult to understand and to respond to. In the thick of the regulatory whirl in the Northeast was something called the Ozone Transport Commission.

The Ozone Transort Commission, as one environmental attorney put it, "was the single wild card of the Clean Air Act Amendments of 1990." The wild card identified a region, the Ozone Transport Zone, over which the Ozone Transport Commission was given the authority to make megadecisions about the future of atmospheric health. The Ozone Transport Zone stretched from Maine to Virginia. Although the zone varied tremendously in geography, weather, and wealth, its 60 million–plus inhabitants shared something the majority of them seldom thought about, a common air shed. Over the air shed, hot, still, dense masses of air during the spring and summer months routinely created a photochemical soup as unnutritious, as unhealthy, and as ugly to look at as anything southern California could cook up. The main ingredient of the soup was ground-level ozone, or smog.

The Ozone Transport Commission, which consisted of twelve governors and the mayor of Washington, was supposed to do something about that. As directed by the Clean Air Act Amendments of 1990, the commission had until 1994 to come up with a plan.

Like the rest of us, most governors don't know much about the air they breathe. It's just there, thankfully. Did any of the governors know, for instance, that air "is a howling blizzard of bizarre matter and even exotic radiation," as social scientist David Bodanis told readers in the *Smithsonian?* Or that air held "fragments of dead stars, hundreds of DNA-packed fungal spores and radio waves that result from live volcanoes on the Jovian moon Io, along with emanations of almost everything else that's taking place on our planet and far beyond"? Probably not. The people who understood emanations were air control officials, toxicologists, and researchers. These staff people working for governors did the footwork, gathered the data, outlined the economic implications of EVs mandated across the region as a way of keeping the bad emanations to a minimum in a mobility-addicted society.

NESCAUM, the Northeastern States for Coordinated Air Use Management, helped, showing staffers and governors why the California low-emission model, with its ZEV component, would work well in the Northeast. NESCAUM did this by comparing three ways to reduce ground-level ozone in the region. One was reformulated gas, which was already mandated by the federal government across much of the region over the next few years. Second was a federal low-emission plan, which kept ratcheting down on conventional engine technologies. Third was the California model, a schedule of successively cleaner vehicles, reformulated gas, and ZEVs. With all three approaches, initially ground-level ozone levels would drop, NESCAUM said. But after early gains, the total emissions from reformulated gasoline introduced across the region or from the federal plan went up again—a result of projected increases in the number of cars and trucks, which would again overwhelm gains made by new technologies. As NESCAUM's executive director Michael

Bradley put it, "What's killed us in the past, and what's killing us now, is the increase in vehicles."

NESCAUM's science and logic were persuasive. Gradually, despite opposition and lobbying from auto and oil interests, it began to look like the Ozone Transport Commission would, with slight modifications, copy the California program for the Northeast. That meant ZEVs, electric cars, mandated in eleven Northeast states. Two states, New York and Massachusetts, had already passed California-like low-emission plans, well ahead of their sister states. For their efforts, both states were in court, battling the auto industry's political arm, the American Automobile Manufacturers Association, which sought injunctions to stop the mandates.

In Detroit, the prospect of the Northeast mandating EVs in 1998 did not cause any celebrations. A niche market in California was one thing, but electric cars across the whole Northeast was something else altogether.

In meetings with the staffs and governors of the northeastern states, spokesmen from the auto and oil industries pointed out, correctly, that electric cars were not zero emission at all. Utility plants, particularly coal-fired ones like those in Pennsylvania, were bad polluters, and they would provide the power to recharge batteries, simply shifting pollution from tailpipes to the energy source. In addition, cold Northeastern winters would kill batteries deader than stones. You cannot drive a stone. The auto industry argued that cleaner air would come at lower cost, both for car buyers and for car manufacturers, with continued refinements of already very clean gasoline vehicles. Such cars were cheaper to build, easier to sell, and less disruptive to the status quo.

To some environmentalists, this all sounded like an echo: We can't do it. It costs too much. The public doesn't want it. The litany had been heard time and time again, for almost thirty years, whenever the industry faced new safety, emission, or efficiency regulations.

Not so, Big 3 spokesmen insisted. This time is different. Electric cars aren't seat belts, or disc brakes, or catalytic converters, or air bags. Electric cars are new technologies that are not as good as what they want to displace, and at higher costs. Try selling that to consumers.

The auto industry's logic, with its emphasis on incremental change, market forces, and conformity to known patterns of behavior, made sense for the short term, but for the long term the uncertainty and unpredictability of pollution, energy security, public health, and global warming costs were equally, if not more, important. If one looked at things over the long term, the modern world was living it up on the earth's energy savings—a situation roughly equivalent to one generation of a very old family discovering the buried inheritance of the past and deciding to use it fast. The planet's carrying capacity was being overshot.

No matter. The auto industry, supported by the oil industry, held the line at Ozone Transport Commission meetings. America's Big Boys were not going to give in to a mandate that forced technology they didn't monopolize or even particularly like. As an oligopoly, the industry had tremendous power as long as its members, both domestic and foreign, stuck together to defeat regulatory efforts. Until conclusive scientific proof, the ironclad kind the legal system demanded and politicians relied on to take tough, unpopular stands, showed that tailpipe emissions killed and sickened and weakened Americans by the thousands, the industry held a good position—though one that was becoming more and more tenuous. The auto industry (as well as big oil) found itself moving into a situation analogous to the one the tobacco industry found itself in. Cigarettes, long defended by the tobacco industry and its spokesmen, had been steadily losing ground scientifically as evidence of their harm accumulated. There were major differences between tailpipe emissions and cigarette smoke, of course. Auto emissions were more pervasive, were more difficult to follow from source to target tissues, and were the consequence of the addiction of an entire nation, not simply the addictions of those who still lit up.

In the 1990s, scientific proof that ground-level ozone harmed lungs was becoming harder to refute. Asthma, bronchitis, emphysema, and certain types of cancer were being causally linked to smog, air toxics, and particulates, both large ones you saw coughing out of truck exhausts and smaller ones you couldn't see at all. Still, the auto industry chose to marginalize the appeal of electric cars as a reflection of health worries. That the industrialized society caused tremendous health ripples, many of which were laying an inexorable stain around the globe, was an unspoken truth the industry typically denied or did its best to ignore. There were exceptions, such as Volvo in Sweden and PSA Peugeot-Citroen in France, as well as individuals within the auto companies who pushed for a more enlightened approach, but they were the exception. In America the ongoing strategy of the Big 3 was to attack EVs and their technologies while reminding Americans that their way of life, to a large degree, owed its health to gasoline-powered cars.

Yet many Americans were ambivalent about the car and its role in the future. There were many reasons: urban congestion, land-use patterns that were dependent on cars, highway deaths, the alienation of people in machines, air pollution, and so on. Yet talk about how cars were damaging the world was seldom heard from the lips of auto executives. The dark synergies of a continually growing, continually expanding auto industry were not extrapolated, at least for the public's view. The industry's central goal continued to be to sell more cars everywhere. There might be 500 million people starving to death in India, as one automotive magazine publisher told an audience of industry analysts at the 1994 Detroit Auto Show, but there were also 300 million middle-class Indians who could afford cars, and the industry wanted to sell to them. The solid front a united industry presented asserted that cars are good, gasoline-powered ones are best, regulations are anti-American, and the marketplace must rule. No rebuttal: end of case.

The Plot Thickens Like Smog

When the final vote on the Ozone Transport Commission petition was taken at the Omni Shoreham Hotel in Washington on February 1, 1994, lobbyists for auto and oil interests were out in force. They were trying to convince the governors to reconsider. A straw poll had shown that the petition would pass, either eight to five or nine to four. Environmentalists from the Sierra Club, the Conservation Law Foundation, the Union for Concerned Scientists, and others, together with health officials and representatives from the American Lung Association, urged the governors and mayor to stand fast.

During the final, frantic hour of lobbying before the vote, Daniel Greenbaum, Massachusetts's commissioner of the Department of Environmental Protection, decided he had to take a break. He had to get away from the mostly fruitless efforts.

Greenbaum made his way to his hotel room, turned on his TV, and channel-surfed. A liberal with a master's degree in city planning and a propensity for musing about what he called "the larger architecture of how we do clean air in America," Greenbaum did not regularly watch TV. Now, sitting before the tube, unwinding from months of effort on behalf of cleaner air, he stopped on a Buick ad.

A concerned, anxious husband was driving fast to pick up his beautiful wife. The ad's message was: air bags—aren't you glad you have them?

Now, aren't these the same people who told us no one would want air bags? Greenbaum asked himself. Aren't they the ones who said the public would reject them, that they cost too much, that they wouldn't save lives?

Throughout the months of negotiations with the Big 3, Greenbaum had been amazed at the different strategies each of them brought to the table. The companies had revealed themselves to be resistant to the new, afraid to go boldly into unfamiliar areas, and scared of rather than attracted by clean-car technologies. Of course, it could have been an act, just to kill the petition, which sustained the California mandate—but if so, it was a good act. In addition, Greenbaum knew, the outcome of the whole prolonged, tedious process was now shadowed by a new development. Last September, on the south lawn of the White House, with the CEOs of the Big 3 on hand, President Clinton had announced the formation of the Supercar Project (the project soon became known as the Partnership for a New Generation of Vehicles). After mentioning an old Ford Mustang he'd rebuilt when he was young, President Clinton said that a billion dollars was going to be spent on the Supercar, a project that would alter the transportation landscape by producing a prototype that got eighty miles per gallon by the year 2003. Ever since that announcement, Greenbaum and other EV supporters had been asking each other why the feds were suddenly cooperating with three multinationals that had shown, time and time again, that they wanted to stifle the new, not encourage it. They hadn't come up with a good answer, though suspicions were that Clinton was throwing a peace offering to big business.

When Greenbaum returned to the ballroom of the Omni Shoreham, the final vote on the Ozone Transport Commission petition was nine to four in favor of the Northeast adopting the California model. New Jersey, despite having the most polluted air in the Northeast, voted no and was joined by New Hampshire, Delaware, and Maryland.

Following this apparent victory, though, environmentalists were edgy. What was really going on, they wondered. Had they focused their limited resources and muscle on what would prove to be a minor skirmish in the larger architecture of how America did clean air—that being the Ozone Transport Commission petition—while the Big Boys had outflanked them straight into the White House?

Thomas Jorling, the feisty and outspoken commissioner of the New York Department of Environmental Conservation, said that the Supercar project "reflects that the energy and focus of the industry is not to move to the car of the 21st century, but rather to oppose the car of the 21st century." California's state senator Tom Hayden, a fellow who actually drove an electric car, a converted Ford Escort, to the capital in Sacramento, compared putting the Big 3 and the feds in charge of America's alternative-vehicle development program to "leaving the brown bears in charge of a salmon hatchery." Mixing his metaphors, Hayden added, "This looms as a Detroit drive-by shooting of California's best hope for clean air and thousands of jobs in a new clean-car industry."

Other environmentalists grumbled that the president's ten-year horizon was too far in the future. They were irked that the status quo power brokers for the Department of Energy, a bureaucracy known for its fiscal appetite and snail-like pace, and the Department of Commerce, which would soon be fighting for its life because of an antigovernment Congress, would be steering the Supercar team. Worst of all was the threat of alternative-vehicle funding drying up. A politically created black hole, the Supercar team of giants might suck green car funding dry if it was outside the interests of the approved circle, or focused on technologies the Big 3 didn't like.

Ron Chapman, chairman of the Partnership for a New Generation of Vehicles Task Force with the U.S. Department of Commerce, said not to worry; the little guys would get a piece of the clean-car pie once it was better defined. But in truth, the environmentalists had been badly outmaneuvered. It was not a killing blow, but there was little doubt that the Big 3 had shown them how the circuitry of power really worked.

In the early months of 1994, the environmental side tried with little success to get sympathetic access to the White House and to the EPA, which ultimately had to rule on the Ozone Transport Commission petition. They tried to get to Vice President Al Gore; they tried to get to Carol Browner, the director of the EPA, whom Gore had brought from Florida to run the agency. They didn't succeed. During this time, under their breath, some environmentalists allied with the EV cause muttered, "Read your book, Al," when attending speeches given by Vice President Gore. They were referring to Gore's best-selling *Earth in the Balance*, a call for long-term environmental sustainability.

Clashing Value Systems

By law, the EPA had the final say on the Ozone Transport Commission petition. But before making a ruling, it had to gather testimony from concerned parties. Three public roundtables were scheduled: the first in Philadelphia on June 8; a second in Durham, New Hampshire, on June 23; and a third in New York City on July 13.

At each roundtable the usual suspects assembled and the usual stances were assumed. One irksome truth the environmental community had to swallow was that since February it had become apparent that some of the governors hadn't really been listening closely to what was going on. They'd been breathing at all those earlier meetings but not really paying attention. But now the auto and oil lobbies had turned up the volume and were making some governors nervous. Supplied with information and data, they were becoming worried about the cost implications of EVs, about things like the infrastructure needs to recharge them, about possible higher gas prices to subsidize them, about political backlash if nobody bought them. Some political backpedaling had begun.

Another thing bothering the EV forces was the worrisome neutrality emanating from EPA officials, most notably from presumed ally Mary Nichols, the assistant administrator for air and radiation at the agency and its highest-ranking official attending the roundtables. In the complex, politically charged, acronym-thick meetings, the two camps laid out their charts, statistics, extrapolations, and emotional arguments. Although the entire California model for cleaner vehicles was on the table, the focus of the disagreements was the ZEV mandate. Auto and oil spokesmen stayed firm: mandates didn't work, EV technologies weren't ready for prime time, and the Big 3 were being pressured into building cars Americans would not buy. They urged the Northeast to adopt their version of a clean car, the "forty-nine-state car," one that burned gasoline, had ultra-low emissions the whole country could benefit from, and could be sold everywhere.

The EV side agreed that a forty-nine-state car and reformulated gas together would reduce ground-level ozone, but not for long. A projected increase in cars would put the region right back where it started from with ozone. Meanwhile, EV developments would languish.

Sentiments within the pro-EV camp were not unanimously in favor of EVs. For instance, the Conservation Law Foundation, simultaneously admired and disdained for its willingness to sue everyone and everything in sight, said the ZEV debate avoided the true problem, which was too many cars using too many resources and perpetuating land-use patterns totally reliant on private vehicles. The Environmental Defense Fund formed its own camp. The environmentalists, though they hoped to break away from a pattern plaguing alliances of the past, not a few of which had fractured acrimoniously before victories were won, couldn't quite achieve harmony. During the roundtables, those in love with clean-car technologies and those who knew this was just a political fight never forged a marriage. Still, the disagreements were not crippling, and the members of the camp did sustain an uneasy truce.

The gap between the environmentalists as a whole and the anti-ZEV parties at the roundtables was much more pronounced. Basically, the two opposing sides were never going to agree.

The clash between them was more than a difference of opinion. It was a clash of worldviews. One side believed that nothing was comparable to the gasoline car, that conservation went against natural, healthy consumer appetites, and that technology, with incremental improvements, would solve future problems as it had those of the past. This was the Houdini approach: we can get out of this, just give us a little time and we'll show you. The other side, at its most extreme, believed that the world was heading toward a ruinous state it couldn't get out of, that America was leading the charge to exhaust the earth's energy savings, and that ZEVs, if not a panacea, were at least a step in a positive direction, one that could help open a green door to an expanded energy consciousness and a more progressive twenty-first century transportation system.

In the words of philosopher-author William Irwin Thompson, the two opposing sides were "thought collectives." No mountain of facts piled high, no dramatic Daniel Webster–ish oratory, no graphs or charts were going to change points of view. That was because facts and arguments can change opinions, but they can't change value systems. It takes chaos, or a crisis, or a prolonged education to do that.

That didn't happen in Philadelphia, in Durham, in New York. The parties went through a ritual. The Clean Air Act Amendments of 1990 had deemed it necessary, but the circumstances were wrong, the biases too deep, the antagonisms too protracted for mutual progress to be made. In this highly paid, overly technical, coolly polite atmosphere, delay, a long-favored tactic of least feasible retreat, was king.

Don't Shoot the Regulators!

The roundtables ended without the EPA making a decision. Other meetings were scheduled. The long-standing pattern, first seen in the wake of Haagen-Smit's work on smog in the early 1950s and again repeatedly in the 1960s, 1970s, and 1980s, as regulators did battle with the auto industry, was still going strong: scientific research had been presented showing, among other things, that ground-level ozone harmed health and that EVs would reduce ground-level ozone; the extent of the harm and the potential of EVs to do anything about it was questioned by auto and oil spokesmen; the automakers moved more boldly into the policy-making process through a combination of testimony from experts and scientists, lobbying of politicians, and a media campaign directed at the public; and delays in reaching any conclusions or initiating any action were the consequence.

Jamie Buchanan, legal counsel for Trudy Coxe, Massachusetts secretary of environmental affairs, was not pleased with what she saw going on. As the meetings resumed, then dragged on, Buchanan doubted the EPA would meet its November 10, 1994, deadline for acting on the petition.

"The auto industry is after delays in implementing what is technologically feasible," Buchanan said. "They're after erosion of zero-emission vehicle mandates because they don't want to eat into their current car market. It's sad to see the EPA, which is supposed to champion the best environmental solution, putting so much into the already rich basket of the Big 3."

In some ways the EPA, consciously or not, had become indispensable to the success of the strategy of least feasible retreat. By gathering more and more conflicting information, by hearing more and more self-serving testimony from both sides, by prolonging its decision making, the agency was ensuring it would not achieve its stated goal, which was to find a path to cleaner air. It was stuck on the old path.

Afraid to leap, the agency was likely to attempt to sidestep its responsibility by seeking a compromise between the parties. A compromise, though, was not the intent of the Clean Air Act Amendments of 1990, Buchanan said. "The act is an excellent example of extraordinary specificity at the statutory level, and the EPA is burping as it swallows that statute because it doesn't go down easily. The EPA is tortuously trying to devise some means around what the Clean Air Act says. But it's not their discretion; it's federal law."

In its defense, the EPA did find itself in a bind. Both sides in the ozone transport debate as much as said they would contest the constitutionality of the petition if the EPA adopted the other side's position. That could mean years of expensive litigation while ground-level ozone levels increased. From the EPA's point of view, a brokered resolution, a compromise, seemed a practical goal, even if it should not have been a legal one.

"The joke is that as long as everyone is unhappy, the EPA is doing its job," said Jason Grumet, a twenty-eight-year-old regulator who had taken over Michael Bradley's job as director for the Northeast States for Coordinated Air Use Management during the ongoing debate. Unlike some cynical observers, Grumet did not view the expensive, tedious, protracted hearing process as "a decision not to make a decision." He saw it, rather, as regulatory masochism. "It's the states asking the EPA to make them do what they could do on their own," Grumet said.

So why didn't the states do it on their own?

Historic precedent said it was hard, even impossible. Attempts to enforce automotive air pollution regulations at the local and state levels, most notably in California, had been expensive, time-consuming, and only marginally effective. One problem was the very nature of mobile emissions. Like Americans themselves, emissions moved around and were hard to keep track of accurately. The emissions varied according to the weather, to driving patterns, to the age of control devices on individual vehicles and how the devices were maintained. But even more important was the fact that cities and states, again with the notable exception of California, had generally lacked the political will to do what needed to be done about air pollution. Jobs were more important than clean air, mobility more important than visibility.

Recall that Congress had not gone for the auto industry's jugular over emissions until 1970, at the height of the environmental movement. The tough, technology-forcing stance Congress adopted had been helped into being by a U.S. Justice Department suit filed against the industry in 1969. The antitrust suit accused the auto industry of having colluded to kill research and development of emission controls since as early as 1953. Tactics used included cross-licensing agreements that eliminated independent players, an internal gag order on all publicity related to air pollution technologies, and a uniform press release refrain: "We can't do it; you don't want it." Settled with a consent judgment (a kind of bargaining agreement in which you don't admit to being guilty but promise to stop doing what you were doing), the suit came at a bad time for Detroit. The industry escaped with a slap on the wrist for two decades of deceit, but Congress had its evidence of collusion. And the stage was set for the 1970 Clean Air Act Amendments.

If the federal government had only fought the auto industry to a compromise and, since the mid-1980s, lost ground in its clean air battles, what chance did states and municipalities have taking the Big Boys to court for control of the atmosphere? Was the job of controlling air pollution beyond the capability of America's institutions? Was such control a contemporary example of the myth of Tantalus, the Phrygian king condemned to stand, thirsty and hungry, in chin-deep water, with fruit-laden branches hanging above his head? Every time Tantalus tried to eat or drink, the fruit receded from his reach. Would the prize of clean air continue to recede, continue to be out of reach of the very institutions assigned its control, whether bills were passed, technologies forced, schedules enforced, and sanctions imposed? Given the history of regulatory control, that conclusion was not hard to draw. That is, America's institutions were incapable of dealing with the scale of its problems.

Still, don't shoot the regulators; it's not their fault. That was the advice of Howard Latin, a legal scholar and professor of law at Rutgers University School of Law. In an article about the Clean Air Act Amendments of 1990, Latin wrote, "Regulatory failure is a complex phenomenon with many causes and manifestations." He put the blame for failure on the regulatory process itself. The process was handcuffed by eight archetypal "laws" of administrative behavior, which he identified. Several of them applied to the EPA's handling of the Ozone Transport Commission petition in 1994. They were as follows:

In conflicts between political considerations and technocratic requirements, politics usually prevails.
Agencies avoid making regulatory decisions that would create severe social or economic dislocation.
Agencies avoid resolving disputed issues unless they can render scientifically credible judgments.
Agency behavior is partly conditioned by manipulative tactics of regulated parties.

Archetypal behavior guaranteed that there would be large gaps between the expectations placed on regulatory agencies, like the EPA, and their actions. Regulators were political appointees, and their agencies political creations at a time when political winds changed direction quickly. Regulators were the messengers the public approved of one year, then wanted shot the next because they brought bad tidings. They had the unenviable job of disrupting the status quo. They were given the authority to do so, but enforcement power was usually weak, and politicians often abandoned regulators when the going got tough. All in all, as Howard Latin pointed out, regulators were ill suited for the tremendous tasks put on their shoulders.

But who, then, could overturn the status quo? In America, the status quo had the power. It had the money. The burden of proof to change it was on those calling for change. In such a paradigm, noted Latin, "Society asks regulators to do impossible things. We ask them to do difficult things under impossible time and resource constraints;

we ask them to behave decisively, selflessly, heroically in ways that are incompatible with normal modes of human behavior."

Then Americans acted disappointed when the actions of regulators so often fell short of their great expectations. "We should not be surprised," said Latin, "for regulatory agencies remain imperfect human institutions and administrators are human beings no better or worse than most."

13

"No Zero-Emission Vehicle Mandate Is a Stake in the Heart"

James Worden

A line of Forces being built for Boston Edison Company at Solectria's facility in Wilmington, August 1992.

We sort of don't back down. If we felt intimidated every turn we made, we wouldn't be doing this.

**—Anita Rajan Worden,
president, Solectria Corporation**

During the summer of 1994, as the Ozone Transport Commission hearings dragged on, Solectria lost confidence that the final outcome of the bureaucratic charade would be favorable to its own success. One hot afternoon in mid-August, a dispirited Arvind Rajan said, "No ZEV mandate is a stake in the heart."

James Worden wasn't quite as pessimistic. "I think the EPA will waffle for a long time over a ruling," he said. "If half the money that's being spent was spent on actually developing vehicles, we'd have twenty good vehicles for sale right now—whether there were mandates or not."

Solectria's president Anita Rajan added, "The fact that the EPA is potentially tied in through the White House to the Big 3—I don't know how we can battle that. Fortunately, the states can decide what they want. Massachusetts is definitely going to stick with the OTC petition."

So it seemed. In Boston, the U.S. Court of Appeals for the First Circuit had recently refused to grant the American Automobile Manufacturers Association an injunction to block the ZEV mandate in Massachusetts.

"Imagine if just Massachusetts had the law," Worden mused. "The Big 3, just in Massachusetts, would have to sell a certain percentage of electric cars." He smiled ever so slightly. "That would get really weird."

Asked what their dominant emotion was about the situation they found themselves in, James Worden and Anita Rajan glanced at each other. She fielded the question. "I think it's excitement," Rajan said. "Often we may sit down and feel intimidated. But that's not for long. We don't sort of back down. If we felt intimidated every turn we made, we wouldn't be doing this."

If the ZEV mandate was killed by the opposition, what effect would that have on Solectria's from-the-ground-up Sunrise project?

"It would slow down a little," Worden replied. "But not much. The support of the utilities and of DARPA and others may adjust a little, but I don't think it's going to go away. It's too late. Electric vehicles are good enough. They're going to get out there. I don't think the California regulations are going to go away. And I don't think it matters too much if the ones here go away or not, because if it gets kicked off anywhere, it's going to get kicked off. Once there are cars being made for California, they'll be made everywhere."

Hybrids were coming on strong, he added. California might consider them as ZEV equivalents, even though hybrids did have very small emissions at the tailpipe when they were burning gasoline or other fossil fuels. A formula could be used to determine how many hybrids equaled an electric vehicle, as far as emissions were concerned.

Politically, hybrids were appealing. But technologically, they could be seen as orphans. Automakers had mixed emotions about hybrids, which still required factory retooling. Electric utilities were dubious about them because they weren't dependent on the grid. And oil companies weren't excited about a car that got the equivalent of eighty miles per gallon. Still, hybrids had staunch supporters. AeroVironment's Paul MacCready was one. "You want a real societal mandate, the hybrid is it," he said.

Amory Lovins of the Rocky Mountain Institute was another. Lovins said a new auto industry could leapfrog the old with a version of the hybrid that he called the "Hypercar." An integrated blend of cutting-edge technologies that reduced weight, recovered energy from braking, cut through air better, and rolled over the ground more efficiently, the Hypercar was a hybrid that didn't accumulate efficiencies so much as multiply them, resulting in extreme efficiencies. On ten gallons of gas, for instance, the Hypercar, theoretically, could cross America.

The Hypercar and other hybrid concepts developed by Lovins and his Rocky Mountain Institute were credited with seeding some of the thinking at the Partnership for a New Generation of Vehicles, the federal–Big 3 alliance that wanted to put an eighty-miles-per-gallon prototype on the road by 2003. Lovin's ideals were inspiring to many futurists, though they also irritated some engineers. A media techno-celebrity of sorts, Lovins might have a genius grant from the MacArthur Foundation, a think tank in Aspen, and a long list of international clients, but he didn't have any real-world experience making cars, the engineers pointed out. For all his talk of an automotive revolution, Lovins also had a rather benign view of the Big Boys. He seemed to see them as large corporations staffed by good people who had lost their way. He prodded them at times, praised them at others. But he said little, at least publicly, about the disruptive and threatening upheaval inherent in revolution, or about the fact that the Big Boys used their formidable power to keep the very agile, clever, smart companies Lovins praised as likely to lead an automotive revolution from even surviving the 1990s.

Mulling over Lovins's work, Anita Rajan said, "A Hypercar is far beyond what we're doing."

Worden said he liked some of Lovins's ideas, and he also liked the idea of a hybrid. But integrating an electric car and a gasoline one under a single hood created complex problems. "It sounds simple. Try building one. It's not as easy as people think."

14

The Sunrise Project and Its Partners

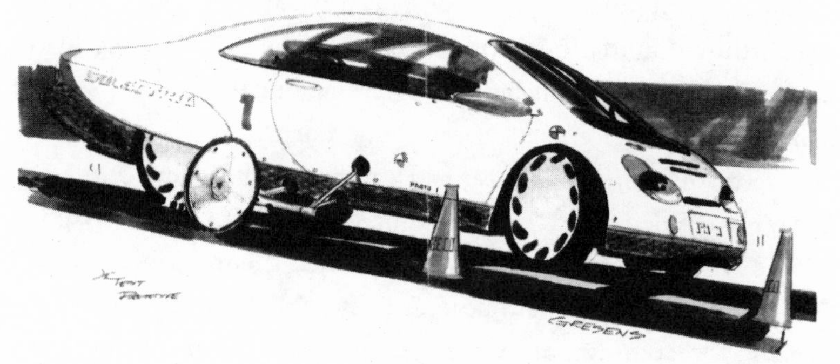

Richard Gresens

This early vision of the Sunrise was drawn by automotive stylist Richard Gresens.

James Worden came to me and said if I raised $300,000, he could do it. He gave me a one-page proposal. It led to a $790,000 grant from DARPA.

—Sheila Lynch, executive director, Northeast Alternative Vehicle Consortia

In August 1994, Solectria was located in a single-story, nondescript brick building in a small industrial park off I-93, about fifteen miles north of Boston. There was no sign, but an orange extension cord, a symbol of sorts of what went on inside, usually snaked out the front door, available to charge cars and trucks. Inside were thirty-five employees. Ed Trembly, now thirty-seven, was no longer the old man of the operation; that dubious honor fell to Mark Kopec, vice president of sales. There was still no gray hair at Solectria, barring a few stray strands here and there, and most of the employees, the majority of whom were engineers, had a somewhat unorthodox look. They wore sandals, and many sported ponytails. A few looked a little rebellious; most seemed restless. To Generation-X visitors, Solectria's offices may have seemed a little like nirvana (even if Worden and his lieutenants bristled at the notion of being lumped with Generation-Xers), what with young, energetic types bustling around with high purpose in an atmosphere of relaxed informality.

On the business side, in two years the upstart start-up had grown from five to thirty-five employees and from sales of roughly $350,000 in 1992 to an estimated $3.5 million in 1994. The company had moved from the cramped shop in Waltham to these larger quarters. To reach the main work area of the new building, a visitor signed in with the receptionist, walked by a fish tank and dusty racing trophies down a short hallway flanked by a couple of private offices, and continued into a warehouse space the size of a small gym. There, hammers banged, torches flared, and guys leaned over pickups and compacts, removing unneeded ICE-age components from Geo Metros and Chevrolet S-10 pickups and replacing them with high-efficiency AC induction motors, electronic transmissions that eliminated the need for a conventional multispeed gearbox, and central controllers, the compact, sophisticated solid-state devices that routed electric signals and had a regenerative braking feature. Stacked high along one wall were lots of wooden boxes that held virtually new, four-cylinder ICEs that had been hauled out of vehicles. Exhaust and emission-control systems, fuel tanks, instrumentation, heating and cooling units—these were removed or replaced as well.

Basically, Solectria was hand rebuilding vehicles, which helped explain the high cost of conversions. For instance, the price for a Solectria E-10 pickup was $43,000; the Chevrolet S-10 it came from cost $10,000 to $12,000 at the dealer. The price of a Solectria Force was $26,000, compared with $8,000 to $9,999 for a Geo Metro at the dealer.

To date, no automaker had agreed to sell Solectria "gliders," which were incomplete vehicles sold at lower prices. A glider for Solectria would be absent the systems mentioned above. Presently, Solectria was trying to negotiate for gliders. If the company could get them from a major automaker, the prices of conversions would drop dramatically.

In a dusty storage alley adjoining the main work area, watched over by a wall of homemade battery-testing devices, was a collection of electric cars. If a cataclysm were to cover northeastern Massachusetts in hot lava all of a sudden, archaeologists digging into the site years hence would find a microcosm of the history of the EV industry in America, circa 1980–1995. Here sat Solectria II, the aluminum car James Worden had built in high school (it would soon end up in a museum). Next to it sat an old French Renault, which had been rusting in a junkyard until the yard's owner saw Worden on television and gave the kid a call. "I got an electric car down here," was all he'd had to say; Worden went down to the junkyard with a new controller and a couple of batteries and drove the car away. Here were the Lightspeed and the Flash, successive generations of a from-the-ground-up electric car awaiting a dubious future: would they become legacies of Solectria's success at forging the way toward a new auto age? Or junk waiting for the next century's James Worden equivalent to see them on interactive video and rescue them from oblivion, as Worden had rescued the Renault? Finally, here was "one of the most successful EVs ever produced," as Worden had boasted to the Subcommittee on Energy of the House Science, Space and Technology Committee in June 1994, when he had testified about the current feasibility of electric cars. It was the Force RS, the record-setting car that Lin Higley had chased during the 1994 American Tour de Sol. Indicative of how fast things were changing in the EV world, the Force RS was thick with dust, its technologies aging, its makers having moved on to other things.

The off-limits inner sanctum of the Sunrise project, adjacent to the historic alley, was a little disappointing. About the size of a three-car garage, the space was practically empty. There was no Sunrise prototype in sight. At the far end of the space stood four guys near an open bay, through which bright sunshine poured. They were all looking up into the air, as though envisioning something. They were the Sunrise team, stylist James Kuo said, after disengaging himself from the group. They were visualizing a spray booth in which they could lay up fiberglass for Sunrise and paint it.

On a shelf sat a small clay model of a seed-shaped sedan, which Kuo identified as the Sunrise. The car's original stylist had been Richard Gresens. Kuo had inherited Gresens's designs and refined them. Kuo's renditions of the car's body parts hung on the wall not far from the clay model. On the cement floor was a yellow foam cutout of the interior. As for robots, computers, and micron-measuring devices, which one would find in most automotive design studios, there were none.

Sunrise was in the plug stage, Kuo explained. A plug was a foam-and-wood car. Wood was used as an aperture on which the foam was mounted, then shaped by hand. The plug was in Rhode Island at TPI Composites, the company that was making a mold from the plug. Once the mold returned, Kuo said, the Sunrise team would use it to shape the full-size production prototype.

The rest of the team wandered over and stood around Kuo. They were model builder Dave Blair and engineers Norm Salmon and Ed Wogulis. Together with project engineer Wayne Kirk, who was in Rhode Island overseeing the making of the mold, this was it, the heralded team behind Sunrise, a car whose mission was to be the most efficient car yet made in America.

Chief Stylist James Kuo

Kuo had just turned thirty. "I'm officially over the hill," he joked. Solectria had hired Kuo the previous December, just after he had graduated from the Art Center College of Design in Pasadena, California, one of the two leading automotive design schools in America (the other was the Center for Creative Studies in Detroit). Before going to the Art Center College of Design, Kuo had earned a degree in mechanical engineering, designed dump trucks, worked as a design intern for Toyota in Tokyo, and raced sports cars, a fact that may have endeared him to Worden.

"Every student's dream was to work for a big car company," Kuo said, recalling his years at the Art Center College of Design. Lately, though, most graduates ended up in Hollywood, creating storyboards and special effects for films, or at toy makers, designing miniature race cars for kids. Jobs at the Big 3 were almost impossible to get; the companies simply weren't hiring.

After nine months with Solectria, Kuo confessed, "I am not electric car guy. It may not be environmentally correct, but I like the smell of fuel."

To date, Kuo's favorite car design of his own had been a police car: bulletproof, bulging with steel muscles, emanating a tough, mysterious look. "I wanted to build a car that was so tough people wouldn't touch it," he said with a smirk. Engineers at Chrysler, which had sponsored the competition for which Kuo did the design, didn't want to touch it either. "They killed the concept; they thought a police car should be friendly."

When Kuo joined Solectria, Sunrise was going to be a small, compact car. James Worden wanted a small vehicle, the size of a Honda Civic. Project partner Boston Edison, on the other hand, wanted a big Lincoln Town Car, or at the very least a full-size American sedan. A compromise had resulted: Sunrise was, as a concept, a 1,600-pound, four-passenger, midsize sedan.

Designing Sunrise had been a hard learning experience for Kuo. Basic engineering principles cherished by Solectria had been in conflict with his sense of style and aesthetics. Compromises had to be forged often. More troubling to Kuo had been his gradual discovery that the engineering mind-set that dominated Solectria didn't respect aesthetics and visual style as much as he felt they should be respected.

The story of the design of any new car is one of compromises—and of making the compromises work, both aesthetically and as engineering. Project engineer Kirk's job was to keep Kuo focused on the aerodynamics and styling of Sunrise while also satisfying the engineers and the major sponsors. "It was a huge effort," Kirk conceded, one that resulted in a constant push and pull, and that left some frustration in its rough wake.

Still, Kuo felt outnumbered. Solectria was an engineering operation, and the engineers didn't know a lot about cars. Kuo had not moved to Boston, with its snowy winters, crazy drivers who passed him on the right, and summer humidity worse than that in Taiwan, where he'd been raised, expecting a Big 3–like setup with a couple designers, three or more modelers, and a fully equipped studio with proper ventilation. But he had, he said, expected more knowledge. In Kuo's estimation, neither James Worden, his chief lieutenants, nor the Sunrise partners knew much about design, manufacturing, or sales.

In a reflective moment, Kuo also admitted that he'd known what he was getting into—or should have known. And it was his own obsessiveness about style that had landed him the job in the first place. Hadn't he, at his interview, sat in a very cool room and sketched his vision of cars for hours, showing the engineers he was as crazy and as obsessive about things as they were?

Modeler Dave Blair was Kuo's ally. Blair, thirty-two, appreciated design. He'd joined Solectria this summer. Before taking the job, he'd been designing aerodynamic windmill blades out of composites. Blair was excited about Sunrise, he compared the complexity of its composite challenges with those of the Stealth bomber. He also hoped the team added some good people real soon or else the project would never get done.

Team engineers Ed Wogulis and Norm Salmon had been students of Mike Seal, a highly respected academic tinkerer at the University of Western Washington, who was good at making oddball prototypes without a lot of money.

As the team waited for the mold of Sunrise to return to the shop in Wilmington in August 1994, Kuo said the project was five months behind schedule. He blamed the lack of knowledge and an unrealistic schedule. "All the people making plans," he said, "they didn't know you had to go through concept sketches, clay modeling, one-quarter size, and full scale. You may, or may not, have to go through wind tunnel tests. We did wind tunnel. The wind tunnel technician called the aerodynamics 'phenomenal.' Finally, there is full-size packaging. All the partners know is jump right into full-scale. It has taken five months to get to that stage. That is why we are five months behind."

With the full-scale plug in Rhode Island, however, Kuo felt some relief. Soon they'd be working on the actual prototype. Getting a prototype finished for EVS-12, the Twelfth International Electric Vehicle Symposium, which would be held in California in December, was still possible. "It doesn't need to be a car that runs," Kuo said. "It just needs to look good."

Making Sunrise look good wasn't going to be easy, he went on. "The goal is still to have a running prototype. But we can't do it with the manpower. From now until December will be hell."

The task Kuo and the Sunrise team faced was formidable. In the bigger picture, of course, a prototype was just a preliminary step, though a crucial one, on the long haul to manufacturing. Whether Sunrise would ever be manufactured was another question. Yet already, money had been earmarked for a manufacturing study, which would begin in the fall. A prototype for which there was the possibility of manufacturing increased the motivation of everyone involved during the early stages. Out ahead of them they could envision something grander, even if it was a long shot.

Now, though, Sunrise was behind schedule. One reason, according to the volatile young engineer Norm Salmon, who soon quit the project, was James Worden. Salmon said he wished Worden would spend more time in Sunrise's inner sanctum. "It's his dream project," Salmon said. "He should be back there more."

Adopting the Right Approach

James Worden had the confidence that makes great things possible, but he also had some lessons to learn: letting things go, praising subordinates, listening to wisdom from beyond Solectria's walls.

Because of Solectria's small size in an industry of giants, it worked from what Rob Wills, technical director of the American Tour de Sol, called "a base of no fear." Neither Worden, Anita Rajan, Arvind Rajan, nor the company as a whole could spare much energy worrying for long about the competition, the mandates, the politics of clean air. Yet to achieve even moderate size, with ongoing thrusts in research, Solectria and its leader, James Worden, had to mature rapidly. "The usual Solectria madness," as Ed Trembly called it, the fine-tuning on the starting line approach, didn't work as the company approached forty employees, faced tough planning choices, and remained privately held.

On the other hand, James Worden possessed traits that made progress, though difficult, assured. He had intense drive. He had a knack for collecting around him the right people to do what he wanted to get done. And he was relatively free of prejudice, which allowed him to touch the truth. Sooner or later, he usually adopted the right approach to the problem at hand. He did this by interacting openly with others, which established a base of understanding. In this way he expanded his personal power. For the leader of a company taking on a nearly impossible task, personal power was absolutely crucial.

Worden clearly demonstrated his freedom from prejudice in his attitude toward, and dealings with, the Big 3. Despite the automakers' sustained efforts to torpedo the ozone transport petition and the ZEV mandates, the loss of which boded badly for Solectria, Worden harbored little bitterness toward them. He maintained communications with Ford and GM. He recognized that their agendas were different than his and that, as he put it, "theirs may not mesh with what's best for the world all the time."

With Sunrise, however, Worden found himself in uncharted territory. His unchallenged authority controlled the project from afar, yet the team needed more flexibility and freedom if it was to learn fast. Because of the lack of experience, which was somewhat counterbalanced by raw talent, the team had to acquire knowledge in its own way. And it did. In a refined version of guerrilla engineering, Kirk, Kuo, Blair, Wogulis, Salmon, and others who later joined the Sunrise team demonstrated how a minimalist approach, with compromises, could effectively serve one overriding goal, vehicle efficiency.

The Major Partners

Solectria might have lacked deep knowledge about the auto industry, but the young company had audacity in spades, along with technological brilliance and an almost uncanny sense of good timing. That James Worden knew what he didn't know—at least most of the time—was crucial, too, though, as just mentioned, he had his blind spots.

For the time being, Solectria also had good partners for Sunrise. They were the Pentagon's Defense Advanced Research Projects Agency (DARPA) and Boston Edison Company, a sort of maverick electric utility whose electric-vehicle program was fueled by three forces: federal regulations mandating utilities to add clean, alternative-fueled vehicles to their fleets to improve air quality; the desire to make money; and James Hogarth. Hogarth, a self-proclaimed EV zealot, headed up the utility's EV program. The two partners were funding the Sunrise project in two phases, channeling the money through Sheila Lynch's Northeast Alternative Vehicle Consortium (NAVC).

Phase one of Sunrise cost $1.3 million. DARPA and Boston Edison were cofunders. Phase one would end when three prototypes—the original, a second to crash test, and a third to use as a manufacturing model—were built in Solectria's low-tech space in Wilmington.

Phase two was the design for manufacturing study. Its purpose was to figure out how, where, and at what cost Solectria and its partners could manufacture Sunrise at low volume. Phase two was just beginning in the fall of 1994 and would continue until 1996 or 1997, ending, it was hoped, with a strong business plan to present to investors. The cost of phase two was estimated as $3.8 million. DARPA had committed money to it, but Boston Edison was having second thoughts about its role in this phase of Sunrise.

Solectria's immaturity, feistiness, and inconsistent business practices irked both James Hogarth and Boston Edison's management. Like many American utilities in the mid-1990s, Boston Edison was also worried about deregulation's impact on its future and was being very conservative in its business decisions. On the other hand, the utility did have positive things to give Solectria besides cash, including a vacant manufacturing space in South Boston, future distribution of the car, and political clout.

In phase two of the Sunrise project, other strategic partners would also be coming aboard. The overriding question was, Could such an unusual alliance usher Sunrise to the verge of production? Was the alliance a harbinger of twenty-first-century cooperativeness of a new kind? Or was it a chimera, an oddball's house of cards, something that would fall apart if the mandates died or if other unanticipated forces suddenly blew through its infrastructure?

DARPA was the key partner. DARPA's program manager for electric and hybrid-electric vehicles was Major Richard Cope, an upbeat, outspoken, and confident marine who, according to the *New York Times,* was "leading the military's charge into electric-vehicle research." With the $68 million Congress invested through DARPA between 1993 and 1995, Major Cope would become sort of a patron saint of EVs.

In addition to Sunrise, DARPA was cofunding other advanced technology programs, often spearheaded by small companies, from Hawaii to New England. Seven consortia, representing more than 200 companies and organizations, had taken shape around the catalyst of Pentagon money. For the companies involved, schedules were tight, people worked hard, and a fundamental high-tech research current was temporarily reversed. Instead of the military-industrial complex conducting secretive, expensive research and development in pursuit of the latest high-tech breakthroughs that might, after they had satisfied defense needs, spin off to the public domain, small, agile clusters of innovators were attempting faster breakthroughs. For instance, the Sacramento Municipal Utility District, a member of the Sacramento Electric Transportation Consortium, developed a small, practical neighborhood vehicle for use on a college campus or a military base. The Hawaii consortium focused on electric buses. EVermont tested cars, pickups, and a four-wheel drive in cold weather to improve thermal management properties. In Georgia, Electronic Power Technology, working with the Southern Coalition for Advanced Transportation, devised a battery-charging system that, in early 1994, helped James Worden set a new twenty-four-hour distance record of 831 miles in a Solectria E-10 pick-up at the Atlanta Motor Speedway.

If successful, the reasoning went, the current of innovation sparked by DARPA would flow from the practical, real world to the exotic, military one, not vice versa, as it had for decades. Breakthroughs would "spin on" to the military through DARPA's largesse, rather than "spin off" after satisfying defense needs.

The rationale behind the Pentagon's change of heart had nothing to do with a love of smallness and everything to do with a post–cold war budget squeeze the Department of Defense had to respond to, at least temporarily. In the past, DARPA had funded experimental projects with a high probability of having no future nonmilitary applications. Occasionally, though, the outcome of such projects was something else altogether. For instance, the agency had developed the original Internet for defense purposes. Some DARPA administrators favored five-year projects, others, ten-year ones. Some narrowed the technology focus, others broadened it. Under Major Cope, quietly and out of the spotlight, DARPA bankrolled EV experiments, at bargain rates.

To some observers, the Sunrise project was an example of the Pentagon's freedom and license to continue spending money where it shouldn't, cold war or no cold war. Did soldiers need Sunrise to cruise the Saudi desert? Not directly. According to DARPA's dual-use paradigm, though, Sunrise's composite skin and advanced drivetrain had potential military spin-ons, as well as commercial applications. Therefore, it was a good investment.

Sheila Lynch, head of the NAVC, swallowed the Pentagon paradox okay. She said, "If several years ago you'd told me I'd be in a room with marine majors and defense contractors and electric utility guys all working on projects—no way! But the Persian Gulf War brought home to the Department of Defense how energy, security, and domestic energy jell." Working with Major Cope and DARPA, Lynch added, had made her grasp a crucial truth: "No one acts faster than the military because they have a battle attitude."

Having spent much of the last decade in the environmental trenches, combating air pollution, traffic congestion, and the dilemma that more and more complicated her life—what to do about the car in a sustainable world—Lynch had a certain battlefield mentality of her own. She characterized herself as "this kind of crazy, environmental-planning nut." But she was charming and smart, and she knew how technology worked. She'd been the driving force behind the Massachusetts Electric Vehicle Demonstration Program, a $2.9 million project that took three years to get through the state bureaucracy and put twenty Solectria Forces in the hands of normal commuter drivers in 1994. At the time it was the largest experiment of its kind to date in America.

Sheila Lynch got along well with Major Cope. The environmentalist and the marine had a mutual admiration society of sorts. The relationship certainly didn't hurt Lynch's chances of receiving money from DARPA ($4.3 million of NAVC's 1993 $11.45 budget was funded by DARPA) and was therefore crucial to Sunrise's development as well. Of NAVC's ongoing projects, Solectria and Sunrise received the largest amounts of cash—not only because Sheila Lynch and Major Cope liked the company, its gene pool, and the Sunrise concept but also because Sunrise was the alternative-vehicle industry's great white hope despite its behind-the-scene problems. As Arvind Rajan put it, "DARPA is really the driving force here. At all the meetings I go to, we keep coming away feeling we're on the ground floor."

Boston Edison Company, a public utility, was more conservative and subdued than either DARPA under Major Cope or NAVC under Sheila Lynch. Nevertheless, James Hogarth, Boston Edison's man in command, was every bit as gung ho about EVs as either of his peers, in some ways even more so. Neither Cope nor Lynch drove an EV to work. Hogarth did, commuting daily in a Solectria Force from Norfolk, twenty-five miles southwest of Boston, to the Prudential Center in the city's downtown. During the winter of 1993–1994, which was especially long, cold, and snowy, he usually drove back and forth without heat to make sure his batteries would make it.

Like Cope, Hogarth had a military streak. He'd graduated from West Point. As a young officer at Fort Hood, Texas, he'd been introduced to the concept of solar stills producing alcohol cheaply and the alcohol being used as a fuel for vehicles. But then "a combat football injury," as he called it, had cut Hogarth's army career short. In 1983 he moved to Boston, joining Boston Edison as a transportation engineer. In 1988 he was one of a select team charged with envisioning the utility's future. Hogarth came away from the experience curious about the possible future role of EVs. By 1991 his interest had grown to the point that he was able to convince his superiors to establish an electric transportation department. Though small—just Hogarth and a clerk—the department had vast company resources to draw from and Hogarth's enthusiasm to fire it. He was a self-proclaimed "zealot about EVs." And zealots, he liked to say, "keep the momentum going."

Hogarth was also histrionic. He liked to make statements like "Electric vehicles can change the world we live in; that's not science fiction," or "This is a made-for-TV movie," when expounding on one of his favorite subthemes: conspiracies to thwart EV progress. At the center of the conspiracies he usually positioned the largest industry in the world, led by the Big 3.

James Hogarth didn't much like the American automotive industry, but he didn't blame it for what he saw as a dismal state of industrial affairs in the country. "From my perspective," he said, "the entire American industrial complex suffers from looking at bottom-line results today as opposed to investing in the future. I wouldn't call it unethical, but I'd certainly call it shortsighted."

Sheila Lynch characterized Hogarth as a man "with grandiose schemes." Hogarth's biggest scheme, the one that most possessed his considerable imagination, was Sunrise. He imagined Sunrise being built in Massachusetts and sold around the world—if not next year, as Sunrise stylist James Kuo said, then at least by the year after. Toward that end, Hogarth had convinced Boston Edison's leadership to agree to distribute Sunrise if and when it was mass-produced—but not as a compact. "Those two-seat rinky-dinky cars don't hack it in this country," Hogarth said. "Sunrise is an opportunity to offer a luxury-size car interior on a compact's wheelbase."

In the fall of 1994 that was what the entire Sunrise team wrestled with in Wilmington: fitting a midsize interior into a compact's exterior. And although James Hogarth's expectations continually struck Kuo as unrealistic, the stylist had to admit that the mercurial, imaginative fellow pushed the project as hard as anyone. It was bothersome to everyone at Solectria, however, that Hogarth's enthusiasm sometimes blinded him to Sunrise's complexity and problems.

That Major Richard Cope, Sheila Lynch, and James Hogarth headed their respective organizations simultaneously was a positive twist of fate for Solectria. In a more traditional and scrutinized environment, Sunrise might never have happened. As things were, Sunrise became the prism through which the Pentagon, Boston Edison, and the NAVC collectively focused long-range intentions without having to get into bed with one another. The partners worked well together, as they discovered, although project engineer Kirk often found himself balancing strong personalities and conflicting opinions to keep the program on track. Ultimately, all the players wanted to see their dreams manifested in a futuristic clean car. If they succeeded with Sunrise, it would be important to the EV industry, and to the greening of cars in general, not because the car was likely to be manufactured in large volume, though it might be, or because Sunrise might be bought by thousands of early adopters, which also could occur, but because the oddball alliance demonstrated how a new kind of cooperation could overcome the status quo with speed, ingenuity, and idealism.

Smuggler's Notch Meeting: October 1994

The Defense Advanced Research Projects Agency held meetings for its consortia every three or four months. Representatives from each consortium updated their peers on the progress being made in battery, material, and drivetrain technologies. There were sessions on public policy, on demonstration projects, on commercialization prospects. There was lots of networking. Each consortium took its turn hosting a meeting. In October 1994, the event took place at Smuggler's Notch, a resort near Mount Mansfield, Vermont, during the height of the foliage season.

A number of attendees came straight north to Vermont from Providence, Rhode Island, after attending the sixth annual Sustainable Transportation Symposium and Trade Show. Speakers had included New York Power Authority CEO David Freeman, who lambasted America's ignorance about its illusory energy abundance; former General Motors CEO Robert Stempel, who reviewed the technology evolution of the Impact; and Rocky Mountain Institute metaphor spinner extraordinaire Amory Lovins, who explained how the Hypercar could capture the snowballing of mass savings in an ultralight hybrid—if only someone would make it.

At Smuggler's Notch on the opening morning of the DARPA meeting, Major Richard Cope told his audience, "Don't get the idea

The DARPA meeting in Vermont didn't impress McAlwee. Echoing the comments of former Solectria engineer Norm Salmon, whose volatility he seemed to share, McAlwee confided that the level of knowledge here was low. "A lot of people are on the steep part of the learning curve," he said. "There's a shortage of wisdom and an excess of information. Generation-X has scoped out the situation. But we don't have the money. The older crowd has a certain fatalism forming their consciousness. But they're in control."

Solectria's Generation-X contingent was conspicuously represented. Not only James Worden, but Anita and Arvind Rajan and Sunrise project manager Wayne Kirk were there, along with several other employees. On the second day, Worden, wearing a suit and tie, gave a presentation on AC-induction motors, emphasizing the systems approach to efficiency of operation as the best path to follow. It wasn't crucial to have the most efficient motor, controller, transmission, or aerodynamics, he said. What was crucial was to have the most efficient total vehicle system.

The systems approach was vintage James Worden. "I'm very much a systems person, even though I work with components," he explained later. "It's good to have hold of the whole system. I usually have several projects going at the same time. Sometimes they're related; sometimes not." At MIT, he had been taught the systems appproach, one that tried to balance efficiency, costs, and interactions. "Every system is affected in some way by every other system," he added. "You don't want to bump one system into a less advantageous position; that can screw everything up."

This partially explained James Kuo's lament that Worden didn't care about the look of Sunrise. An overemphasis on looks, according to Worden's approach, might unbalance the larger whole. Worden, though, had not made his systems thinking clear to Kuo, nor had Wayne Kirk, whose job it was to buffer and integrate divergent opinions, managed to get the two men to see eye to eye on the looks versus efficiency equation. Worden apparently never said to Kuo, as he did in his office one afternoon not long after the Vermont meeting, "We use system thinking because we know how much of a pain in the ass it is if things aren't right, or you sacrifice one thing for another."

On the third day of the DARPA meeting, there were several provocative but sparsely attended presentations on safety, and how ignoring safety could threaten the entire EV industry. For instance, all the industry needed was for James Worden to have died from inhaling bromine gas in Phoenix in 1992. Or someone burned horribly in a battery fire, such as those that had plagued the Ford Ecostars the past summer, prompting a total recall of the fifty or so made to date. Or the failure of a poorly designed component, such as brakes, resulting in a vehicle getting squashed by a cement truck, which had almost happened at the 1994 American Tour de Sol. TNE II, a 400-pound solar racer made of Kevlar and carbon fiber, had crossed an intersection in front of a cement truck barreling down on it, cut an arc across a lawn, missed a tree, and crashed into a spring tile jutting out of the grass. The brakes had failed. Luckily, the driver only gashed her knee.

On the final morning of the meeting a race of consortia pickups was scheduled for the steep and winding road leading up to Smuggler's Notch. The sun came out. Tourists, locals, and some media assembled. They inspected, touched, and slid into some of the display vehicles. The hybrid M113 armored personnel carrier, which had a 275-horsepower AC motor, lots of batteries, and a Chevrolet 454-horsepower gasoline engine, made fast and relatively silent sprints up a ski trail in the background, its treads clattering, much to the delight of kids who got to ride as passengers. Once the half dozen pickups drove off in the race, the Horlacher broke down by a Vermont farm. The farmer strolled out and spotted the problem right off—a busted engine mount. He got some tools from the barn, helped fix the Swiss green car, and with his wife and three kids watched it drive away.

Back in Wilmington

Back in Wilmington, no one was driving the first Sunrise yet. But, surprisingly, the project was almost back on schedule. The mold had arrived from Rhode Island. New team members had been hired. Most days now, ten or so team members in white smocks worked on the prototype in the cramped quarters, intent on getting it ready for the Twelfth International Electric Vehicle Symposium (EVS-12), which would be held at Disneyland in early December.

One evening, after a game of ultimate Frisbee with his coworkers, modeler Dave Blair said the process by which Sunrise had suddenly come together amazed him. Smiling enigmatically, Blair declared, "The whole thing is organic, like a tree."

On the cement floor the mold looked more like a huge pumpkin; it could have been Cinderella's carriage after midnight. Sunrise was a fairy-tale project in many ways. It was a funky, handcrafted symbol of a new automotive order struggling to come into existence—a 1,600-pound symbol being built by a small team struggling up a steep learning curve toward an idealistic goal, one that conventional thinking said couldn't be reached quickly, or possibly at all. To work long hours with limited knowledge in the shadow of such doubt took nerve and tenacity, but the goal for Sunrise was clear. Ultimately, in the terminology made popular by futurist Amory Lovins, Solectria wanted to use Sunrise to leapfrog into the future. The company wanted to be a star in what Lovins called the car's next great adventure.

With Sunrise, Solectria was attempting more things Amory Lovins talked about than any other company in America: eliminating weight, refining aerodynamics, capturing the "technological compounding" of energy savings that a multifaceted approach to an automobile of the future generated. But Solectria shunned hybrid EVs. Worden didn't think their sophisticated complexity was needed now, that it would most likely delay the success that simple, battery-powered electric cars offered immediately, keeping cleaner cars on the distant horizon rather than putting them on the roads soon.

California was softening on hybrids as alternatives to ZEVs to comply with the 1998 mandate. And some automakers were working on hybrids with enthusiasm. Compared with EVs, the hybrids promised better range and performance. Manufacturing-wise, they had fewer transition risks because they relied more on ICE technologies. Politically, they were more palatable because they weren't completely divorced from oil. Volvo, Mitsubishi, BMW, Volkswagen, Audi, and Chrysler all had hybrid projects in various stages of development.

Whether a prototype was an EV, a hybrid, or some other clean-fuel option, such as a compressed natural gas vehicle, getting the weight down was a common goal. Aluminum was one way to do it, but an expensive way. Ultralight steel was also an option, one the steel industry was spending millions promoting. But composites, because they were not only lighter than the newest generations of steel and aluminum skins but also significantly stronger, held out the greatest potential.

Structurally, composites are simple. Plywood is a composite, as are reinforced concrete, fiberglass, and wattle and daub. The word "composite" means two materials made into one material. Fibers in a resin matrix are what most people picture when they think of a composite. The fibers can be woven or knitted together in different shapes. The weave or knit affects a composite's strength, resistance to damage, energy-absorption characteristics, and willingness to conform during the manufacturing process to complex shape demands. When fibers and resin matrices are laid in a mold, they undergo a chemical reaction, called cross-linking, and harden into the shape of the mold. The types of fibers and resin matrices determine the manufacturing process. Processes for more sophisticated composites can be very expensive.

Sunrise prototype number one (P1) was not an expensive composite. It was old-fashioned fiberglass laid up with a mix of marine, aerospace, and automotive techniques inside the orange blob mold, which was the shape of the exterior of the car. The car was eventually pulled from the mold in two big sections: the top and the bottom. Inserts went where doors and windows would go later.

Occasionally, reporters were led into the dusty, fume-rich inner sanctum by Arvind Rajan, who, in addition to his other roles at Solectria, was also corporate press liaison. The reporters took notes about the sketches taped to the wall, were shown various door options—one in carbon fiber, even one in wood—adjacent to the sketches. They asked James Kuo, who was now focused on the prototype's interior, a few questions.

As summer shifted into fall, Solectria unofficially initiated phase two of Sunrise, the design for manufacturing study, the goal of which was to identify ways to build affordable Sunrises in low volumes. Other partners, including Charles Draper Laboratory, Dow-United Technologies, and Arthur D. Little, Inc., tentatively came on board to work on composites, manufacturing processes, recycling, and other challenges of a car-making start-up. Eventually, of course, Solectria was going to need big investors if Sunrise was going to be manufactured. And finding such investors was going to be tough, since car manufacturing was astronomically expensive, at least in its conventional form, and certainly would not be cheap regardless of the shape it assumed. Boston Edison's management had made it clear to Hogarth that the utility did not consider itself an investment angel. It was a truth, the irrepressible James Hogarth said, "that makes it hard for me to convince other companies that they ought to invest in Solectria."

Most investors, whether blessed with halos or clothed in more conventional attire, wanted to hear about markets, product fit, and potential sales. Electric vehicles did not have a market yet. They had a mandate, one that might disappear. No one—no Big Boy, no small guy—was yet manufacturing for the general public but only for fleets forced to buy EVs by clean air regulations. Higher-volume sales, which would be necessary if manufacturing economy of scale was to be achieved, remained contingent on the ZEV mandates in California, New York, Massachusetts, and, possibly, across the Northeast—if the EPA could ever make the ruling it kept putting off.

In this risky environment, with its multiple uncertainties, the amount of money an EV manufacturing operation would need to get started and to survive—if the market was there and bought vehicles in volume—was difficult to estimate. Would an assembly-type operation the size of a regional soft drink bottling plant do the trick, as Amory Lovins predicted? Should a manufacturer make certain pieces of the car? How much would transportation of components add to the price if the components all came from scattered locations to a small assembly point? How did various manufacturing scenarios play out environmentally? After all, a ZEV company ought to sustain its environmental awareness throughout its operations—or could it?

For Solectria, answers to all these questions were important, tough, and contingent on many factors beyond the company's control. In addition, funding from its most loyal partner to date, the Pentagon through DARPA, could be counted on to dry up after the study for manufacturing was complete.

If that wasn't enough to worry about, there was also a political sea change brewing in Washington, one that would soon bring fiscal conservatives into power and threaten to end federal funding for high-tech transportation research and development. That October, Solectria got a taste of the imminent change. The company was invited to participate in the first-year celebration party of the Supercar, a.k.a. the Clean Car, and the PNGV, or Partnership for a New Generation of Vehicles. Precelebration press materials boasted that with a billion dollars the project was going to "reinvent" the car. The new car, the Supercar, would triple the range of the typical 1994 American sedan, getting eighty miles per gallon rather than twenty-seven, which was the average. As part of the techy edge of the event, Solectria provided United Technologies, a flywheel manufacturer, with an E-10 pickup. The pickup was trucked to Washington, D.C., where President Clinton, Vice President Gore, and the CEOs of the Big 3 were supposed to rendezvous on the White House lawn. At the last minute, however, the flywheel-powered pickup was "uninvited" to the high-profile event. The pickup remained on the other shore of the Potomac River. GM's CEO Jack Smith suddenly excused himself from the event as well. Smith was fuming over a just announced, possibly huge recall of his company's C/K pickups, whose sidesaddle gas tanks occasionally exploded in accidents. Chrysler's top executives did not attend either; they were in China, pitching their version of the "people's car" to the government. Ford CEO Alex Trotman did make the party, but the orchestrated affair came off with a limp rather than with a high kick celebrating the alliance of old adversaries. It seemed they were still adversaries in more ways than they wanted to admit publicly.

That October, the EPA announced its decision on the Ozone Transport Commission petition. The EPA extended its November 10 deadline for making a ruling, just as Jamie Buchanan, legal counsel for Trudy Coxe, Massachusetts secretary of environmental affairs, had suspected it would. The agency said it would keep holding meetings. Behind the scenes it promised to keep trying to forge a compromise of some sorts between the opposing sides in the imbroglio.

The one-truck recall, the delayed Ozone Transport Commission vote—both went against Solectria's, and Sunrise's, forward momentum. True believers, however, were not daunted by trivialities like political showmanship and the bureaucratic sidestep. If they were, they would quit doing whatever it was they were doing and find less risky work. True believers were attracted by impossible odds. They kept the momentum going any way they could. In Wilmington, the Sunrise team just kept working on their car.

15

Anaheim, 1994: Twelfth International Electric Vehicle Symposium

Printed by permission of Electric Vehicle Association of the Americas

The World Electric Vehicle Association's twelfth biennial meeting drew over a thousand engineers, policy makers, executives, and a few loonies to the original Disneyland in Anaheim, California, in December 1994.

Without dynamic and massive change, the automobile companies could well join the ranks of the industrial dinosaurs.

—Sheldon Weinig,
keynote speech, EVS-12

As one entered Los Angeles from the air, the city looked like a colossal salute to automobility. Scattered subdivisions clung to hills, commercial and industrial areas were neatly circumscribed by ribbons of asphalt. In the air there was only a mild haze: Santa Ana winds had blown most of the smog out to sea.

The World Electric Vehicle Association's twelfth biennial meeting officially opened at the original Disneyland on Walt Disney's birthday, December 5. Over a thousand of the world's EV elite, including engineers, policy makers, executives, and a few scattered loonies, the largest gathering of them ever, had come to the "Happiest Place in the World," as Disneyland billed itself. Yet, as speakers noted, journalists joked, and attendees murmured between the sessions on technology, public policy, politics, demonstration projects, and so on, the prevailing question was this: Is the EV industry in Frontier Land, in

Fantasy Land, in Future Land, or—worst of all—in Never-Never Land?

Press days and a weekend of displays of thirty-plus EVs for the public to inspect, ask questions about, and ponder preceded the official opening of EVS-12. There was an EV parade down Main Street USA. Majorettes danced; confetti rained down from rooftop blowers like red, blue, green, and yellow leaves; hidden speakers blared:

> Living it up in America
> Living it up on a new highway

Even the Big 3 came to EVS-12. General Motors brought about a dozen-plus Impacts, including the multidecaled, modified, bright orange Impact 320, which had set a land speed record of over 183 miles per hour in March 1994. Ford brought several Ecostars, Chrysler, its sluggish TEVan.

A Japanese contingent came en masse: Mazda with a Miata on NiCads; Nissan with a two-passenger station wagon called the Avenir EV; Toyota with its EV-50 city cars.

The French came in style. A team consisting of Peugeot, SAFT battery company, Electricitie de France, the city of Paris, and smaller suppliers treated EVS-12 as a major auto show. They had the stylish graphics, a classy display, hostesses, wine and cheese, a comfort salon with cocktails. They also announced a captivating plan, called the "French strategy," for putting EVs on French roads in 1995. It was a team plan anchored in practical measures—battery leasing as opposed to outright purchase, buyer incentives to lower customer prices, a network of charging stations in Paris and other cities, and free parking for EVs in congested downtowns. What was just as remarkable was the fact the French said without equivocation that too many vehicles and their air and noise pollution were ruining the quality of their cities, and EVs were a way to do something about it.

Filling the lesser ranks at EVS-12 were Renaissance's Bob Beaumont, the veteran EV maker, whose two-seat Tropica sportscar was still trying to launch itself into the Florida marketplace; U.S. Electricar's urban delivery van, which the company said it planned to sell in Mexico City; Electric Fuel Corporation's zinc-air battery system, which was scheduled to be used by Deutsche Bundesposte Postdienst, the German Federal Postal Authority, in short-range electric vehicles after a field test scheduled for 1995; the Solectria Sunrise; and other small-guy exhibits of vehicles, battery alternatives, and new technologies.

In such a competitive atmosphere, audacity counts, especially when you're small and want to do something big. When Solectria unveiled Sunrise, which didn't run and had to be steered by someone kneeling down and muscling the front tires, beneath a tent in front of the Disneyland Hotel, and announced it could build 20,000 of them in 1997 for $20,000 each, the company was being audacious. After all, GM, which had Impacts parked and charging at a kiosk that suggested a futuristic EV refueling scene, said it couldn't be done. Mobil Oil was running a series of ads challenging the real costs of electric cars. One recent ad had claimed it cost $10,000 to $20,000 more to make an EV than to make the equivalent gasoline car—extra money, Mobil said, that got drivers poorer performance, limited range, and questionable emission benefits.

No matter. The media still made Sunrise the darling of the symposium. The Associated Press ran a photo—Sunrise aloft, backed by clear skies, promising to generate "sparks of rivalry" in an emerging electric car industry—and text highlighting the manufacturing boast. The prototype was featured in *U.S.A. Today.* CNN showed up and interviewed James Worden in front of the yellow-and-brown prototype with the logo that James Kuo admitted having knocked off in a couple hours.

Worden, the suited Generation-X CEO, had a dry cough and looked tired. Before facing the camera, he had shown a little anxiety about all the attention Sunrise was getting. Not that he didn't want the car to attract lots of attention, but he was bothered by the fact that the prototype, which was a concept for a distant and uncertain future, was upstaging two Solectria mainstays to either side of it: a purple 1995 Force and a red E-10 pickup. The conversions were real cars you could buy now, Worden kept emphasizing to those who would listen.

To some observers at EVS-12, the Sunrise prototype suggested the next-generation Impact. But that was an exterior judgment only. Sunrise had only air beneath its hood and no batteries in its battery compartment. The Impact's performance, road feel, and crashworthiness all were demonstrable; in fact, GM had Impacts taking press and symposium delegates on ride-and-drives throughout the duration of the symposium.

The typical response to a drive in the Impact was one word: "Wow!" From the ergonomics to the instrumentation to the acceleration, with its straight-torque line that no ICE vehicle could duplicate, the Impact felt and handled like a luxury sports car. After driving one, it was easier to understand why GM said it was having a hard time getting demonstration cars back from PrEView Impact drivers. PrEView Impact was the twelve-city, two-year test GM had kicked off the previous June. Selected drivers got a car for two to four weeks. A utility partner installed a charger in each driver's garage. Drivers kept logs, and an onboard black box, similar to a flight recorder, gathered data. Although preliminary reports on the program had been strong, with 83 percent of the drivers saying the Impact met his or her needs, GM downplayed them.

Given the disparity between what the Impact was—a high-quality, fast, performance-focused vehicle—and Sunrise could only hope, with millions of dollars and months of work, to become, why was the latter accorded so much attention not only by the popular media, which typically went running to David and Goliath stories, but also by some members of the automotive press, who attended in significant numbers? Before elucidating the reasons, it is imperative to state that some members of the auto press sided 100 percent with Sean McNamara, marketing manager for GM's EV program. McNamara said, "If we did what Solectria did, the press would laugh at us." Some of the auto reporters' comments were:

"Get real!"

"The car doesn't even roll!"

"It's one off, for Christ sakes!"

So why didn't all of the press laugh at Solectria? Scale, for one thing. Knowledge of just how close the EV industry was coming to actually being a force in the auto world for another. Intellectual excitement for a third.

Scale changes everything. Small-scale Solectria, with the big plans announced at EVS-12, got the lion's share of attention because the hazy plans were what the audience longed to hear: a little company was going to leap into the complexities of manufacturing. The idea was a bit intoxicating; if achieved, it would validate ideas planted in many delegates' and journalists' minds by the writings of Amory Lovins.

Knowledge that technologically the EV was ready to be launched was palpable at EVS-12. No European or Japanese carmaker had a display vehicle that rivaled GM's Impact, but the potential was there. The electronics, components, and charging systems necessary to build EVs were all on exhibit. The challenge, as several speakers emphasized, was now one of getting the infrastructure built and the vehicles effectively marketed. As for Sunrise not being a roller, GM and all the other major automakers occasionally unveiled concept cars that didn't run; in fact, that was the reason Solectria had gone ahead and done it here.

The intellectual excitement in evidence throughout the symposium gave pause even to cynical journalists. True, you didn't see EVs in the streets yet. True, the manufacturing costs would initially be higher than for gasoline cars, just as the Mobil ads said. Yet in only a couple of years the technologies to make EVs feasible had improved manyfold. Environmentally, the futuristic vehicles held out the promise to clean air, to reverse the globe's cancerous atmospheric deterioration, and to help put the brakes on global warming, an issue rapidly gaining worldwide attention.

Ken Baker, GM's vice president of research and development, presented a wish list at the symposium. Baker said GM would like to see a charging infrastructure being built across America, guaranteed large-volume buys of EVs by government fleets, government incentives to cut buyers' costs, and more government money spent on battery research and on public recharging stations. With those changes in place, Baker said the EV industry would have a chance to survive.

More sanguine, California's Senator Barbara Boxer said that the biggest hurdle facing EVs "is old-fashioned human skepticism." And the best way to combat the skepticism about EVs was to demystify the technologies, to get the vehicles out on the roads. Thousands of kids wrote her every day about the environment, Boxer added. The kids were worried. "If kids can't breathe," Boxer said, "they don't have a very good future."

The most enthusiastic endorsement of Solectria's Sunrise project, as far-fetched and impossible as it still seemed even when buoyed up by media hyperbole and great hope poised on the weak trusses of intellectual excitement, came in an oblique way. Talking about future challenges to a large audience in the grand ballroom of the Disneyland Hotel, Sheldon Weinig, founder and president of Materials Research, a division of Sony Corporation, recounted the lessons to be learned from the explosive growth of the semiconductor and computer industries.

The lessons were that giant companies learn too slowly to lead paradigm shifts, and problems looking for solutions accelerate change rapidly. Large companies, including General Electric, Westinghouse, RCA, and Sylvania, had all attempted to enter the microprocessor race in the beginning, Weinig recalled. But they'd dropped out, and new companies such as Intel, Advanced Micro Devices, National Semiconductors, and Texas Instruments had prevailed. The big companies that tried to but could not make the transition to semiconductors from contiguous bases of operation failed, Weinig said, "because they were unwilling to—or perhaps incapable of—embracing the new technologies. They lacked the necessary management style that required hands-on, risk-taking, technical avant-garde, unorthodox approaches to problem solving."

Even if the Big 3 did manage to embrace the new technologies and to transform their style, which they hadn't done, they had to respond to new customer demands, Weinig added. "Customer demands today are almost unbelievably different than they were even ten years ago." Raising his voice, Weinig warned that consumers today were more assertive and louder, "and they wield a very big stick. Without dynamic and massive change," he went on, "the automobile companies could well join the ranks of the industrial dinosaurs."

Weinig left the audience with the impression that the auto industry was so slow, so stuck in dated practices, so committed to advancement of laggards from within rather than bringing in fresh people from without that it was but a matter of time—a decade, maybe sooner—before the oligopoly imploded from its own weaknesses. Yet the Big 3 had just gotten off the doormat. Since 1992, profits had improved, labor and management fat had been trimmed, and quality had improved. Tens of thousands of hours had been spent on reeducation and leadership, on customer awareness, sensitivity training, and better stewardship practices. Teamwork had become the mantra of the contemporary enlightened American corporation. So, were Sheldon Weinig's comparisons between the giants who didn't make it in semiconductors and the Big 3 worth his fee? Or had he simply titillated the EV elite during their final day at Disneyland?

Paul MacCready, for one, didn't buy Weinig's analogies. "When you're dealing with moving people—that is, the whole body, not just moving the mind like the computer does—the wide-bodied jet business is the analog. If you want to start another Boeing in your garage, more power to you. But I don't think you're going to succeed." A lot of small companies hustling for an advantageous position in the EV business saw things differently, MacCready acknowledged, convinced as they were that they were about to topple the inertia-ridden giants. "They see that Apple Computer got started in a garage, and look what it did to IBM. But I think we misinterpret this."

The real opportunities, in MacCready's mind, were for small companies smart enough to align themselves with the Big Boys rather than go it alone. "It's the only way you're going to make a big effect," he said during EVS-12, an event he decided not to attend. Many of the small companies and their leaders had been irritating MacCready as of late. They didn't listen, he said, they didn't think, they were at each other's throats too often. He'd stopped making presentations at seminars because those attending didn't want to listen to the realities of batteries and to the limits now severely hampering cars powered solely by them. Those limits were the reason he continued to advocate hybrids as a more feasible alternative to the purely gasoline-powered vehicle of today.

As for Solectria, MacCready said he didn't think that trying to become a car-making company was a very good long-term strategy. "As long as there are regulations and political drivers, there will be opportunities for companies like Solectria," he went on. "But the big car companies are trying to deal with real solutions and don't make a lot of noise about it. They don't raise unrealistic expectations. When they really move in with whatever the real solutions are, I don't know what will happen to the little guys. They may get absorbed. There may be good business plays for them. But I think it will be the Big Boys that dominate. The Big Boys are not dragging their heels."

They were not sprinting toward the twenty-first century with clean-car technologies either. As MacCready said, behind the scenes they were doing interesting stuff. But publicly, their strategy was to kill the ZEV mandates, to constantly question the technologies and their appropriateness for the market—all the while keeping progress they were making on the technologies tightly under wraps.

"It's classic automotive technology push," said Sheila Lynch. "Try to stop it. Keep your work in-house, behind closed doors. Someone introduces the product and consumers like it, you better be ready to go with it."

The situation was reminiscent of the 1960s, when the Detroit automakers had colluded to keep emission-control devices off the shelves in California, then hustled to market their own once independent manufacturers made them available. After the independents announced they had developed add-on devices, it had taken Detroit only two months to make a breakthrough after years of insisting it couldn't be done. Detroit automakers said they had developed engine modifications superior to the add-on devices. In a well-documented paper called "Technology-Following, Technology-Forcing, and Collusion in the Auto Industry," W. Alton Jones Foundation senior research fellow Paul Miller traced the history of this tale and concluded, "As a result of their experience, many independent technology developers left the field of emissions control for automobiles, saying that the [process in California] was too unreliable to justify the investments needed to develop cleaner technologies." The 1964 incident was a tip-off that Detroit was colluding, however, and it led, albeit very slowly, to the U.S. Justice Department's suit in 1969, and helped usher in technology forcing.

James After One Drink

At a cocktail party thrown by Boston Edison to celebrate the successful Sunrise unveiling in Anaheim, James Worden was in a cheerful mood.

"Oh, you're getting him after a couple drinks," Anita Rajan protested when a journalist joined the relaxed CEO at a table by the buffet.

"Just one," Worden told her sheepishly.

"How do you feel after the successful introduction of your car?" the journalist asked.

"We're very excited."

"Do you look forward to the next scale you're going to have to shift up to bring it to production?"

"Look forward to it. But it's going to be a ton of work. We need a lot more really good people. We've worked very hard just to get where we are by putting the right people together and to do this carefully. But one thing we're going to do is, we're going to be cautious. I don't want to screw this up. Or let it get out of our hands. Or go away. We're in this for the long run. If the decision is to wait and hold off, or to let it develop a little slower, then we're simply not going to push too fast."

"Jim Hogarth would like to have you manufacturing 20,000 cars, and selling them for $20,000 in a couple years. It sounds great. But is that a workable dynamic—you conservative, Hogarth pushing for fast growth?"

"We're both conservative in a lot of ways," Worden said. "But we're both very crazy in other ways. We both know we have to push the technology to increase excitement and get other people interested. He's very good at that. I think the relationship is pretty healthy. It's annoying here and there . . . to both of us, the pushing and pulling. But we both have a similar dream and it's going to happen."

"Did you expect Sunrise to be the hit of the show?"

"Nope. We thought the Big 3 were going to do some major unveilings here and overshadow everything—with us out in a tent. But it didn't turn out that way."

"Would you, if you had the opportunity, hire some of the GM people working on the Impact to work for you?"

"We really like a lot of the folks at GM. We know that a lot of them in the EV program really want electric vehicles to happen. They're pushing it as hard as they can. They're almost fighting a second war, though. It's enough work to be fighting what we're fighting, with very little money and trying to do things efficiently. And fighting to make the technology better and better. They have not only those problems, but they also have a whole corporate culture that's completely against them."

"So you will be reaching out for Detroit people?"

"Definitely. We'll be taking them on slowly, but we need them."

"Your car has been characterized as the next generation of the Impact—do you agree with that?"

"It's got all the slickness of the Impact. But it's lighter and it's higher technology. With a four-person interior, it's a lot more palatable to the public. We're also keeping it simpler than the Impact in many ways. It has a long way to go. We have work to do now."

16

Sunrise Almost Stalls, Then Rolls, Finally Races

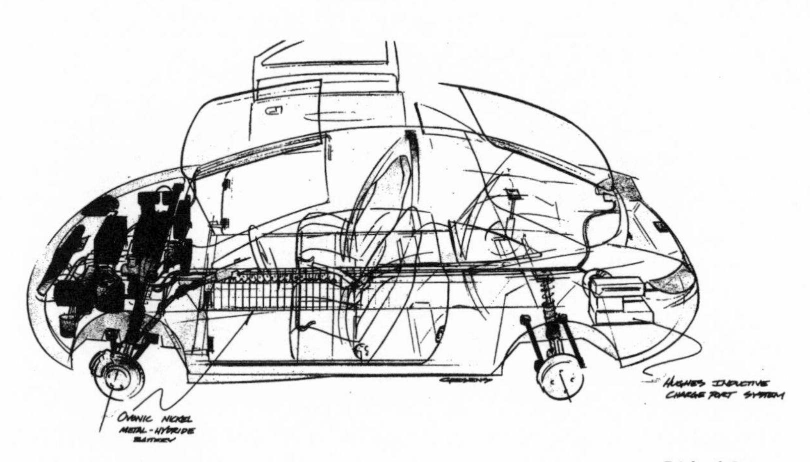

Richard Gresens

The body design and parts breakdown of Sunrise were developed during the design for manufacturing program.

We were just dumb enough to jump in first.

—Ken Sghia-Huges,
Sunrise lead engineer for vehicle safety systems

There was more work to do than even Worden imagined—and a trying period to get through. It began in late January at a meeting Sheila Lynch organized for the partners in Sunrise's phase two, the design for manufacturing study. James Worden, James Hogarth, and representatives of the smaller companies joining the study, which was still just getting off the ground, all attended. Lynch also invited James Womack, coauthor of *The Machine That Changed the World,* to come and talk about agile manufacturing, which might fit into the future of Sunrise. Agile manufacturing, Womack wrote, had been displacing mass production in the Henry Ford assembly-line style, and would continue to do so. Agile manufacturing, as refined by the Japanese and adopted by automakers worldwide, was flexible, lean, and respectful of people; it used less capital, tolerated less inventory, allowed less finger-pointing as a means of avoiding responsibility, and generated less waste. To prep everyone for the meeting, Lynch had bought eleven copies of Womack's book and mailed them out ahead of time.

The meeting, though, was a bust. Despite Lynch's efforts to inform and prepare the participants, two of them, Solectria and

Boston Edison, were not quite ready to deal with agile manufacturing concepts yet. The gap between the cautious James Worden, whose company had built Sunrise, and the zealous James Hogarth, whose company was partial owner of the prototype and the distribution rights, should Sunrise ever reach production, had been widening for months. Now its breadth was painfully obvious.

Hogarth had basically had it with the whiz kids from Wilmington. They might have the best technology, he felt, but they also had the worst attitude. He wanted the company to move faster on Sunrise, to stop pelting Boston Edison and himself with disregard, and to live up to its contracts. "James Worden is afraid to leapfrog into manufacturing because he'll lose too much control to new partners," Hogarth said. "The company is unwilling to work with anything that wasn't invented there."

Worden wanted Hogarth to back off. He wanted the development of Sunrise to move forward slowly, out of the public eye, and not at the expense of Solectria's conversion business. "Jim Hogarth is way off key, trying to blast the thing," Worden said. "He wants to be marketing Sunrise solid, and running it around everywhere. And we're like . . . let's push it back. Let's put it in a back room and develop it. The car we unveiled in Anaheim can't be made, won't be made until it's perfected. It won't be that long before it's perfected. But we're spending a lot of energy on friction."

Control questions, scheduling problems, how to do business—these ongoing hassles had been repressed in order to get the first prototype finished and to Anaheim. Now, though, things were out in the open.

Guest speaker James Womack wasn't impressed. Maybe he'd get involved with the Sunrise project when the likelihood of success was more auspicious, he told Lynch later. Lynch still pressed forward. She wanted the team, despite the sparks, to apply for a National Institute of Science and Technology (NIST) grant for agile manufacturing of composite car bodies. As the person responsible for sustaining Sunrise's momentum and for coordinating the technology partners, Lynch felt that the grant, if they got it, could push Sunrise right to the abyss of manufacturing. The money would pay for the high-risk work of identifying the tools, machinery, and processes for making composite car bodies. Much to Lynch's dismay, the unanimous response of the partners to her grant idea was no.

It was not a word Sheila Lynch liked to hear. After the turndown she stewed for a while. Then she talked to Worden. "I'm not writing a NIST grant myself," she told him.

Worden was adamant; Solectria wasn't going to get involved in this one.

Nothing if not persistent, Lynch attended a NIST meeting to learn how to apply, took home the forms, and talked to Sunrise project manager, Wayne Kirk, and business manager, Mark Docksер. Dockser had replaced Arvind Rajan, who, feeling he had accomplished what he set out to do at Solectria, which was to give it some business shape and two and a half years of his time, had returned to San Francisco. Lynch convinced Kirk and Dockser of the value of the grant and urged them to lean on the recalcitrant Worden to go along with the application.

The grant really did hold out promise, especially given the potential defection of Boston Edison if things weren't ironed out between Worden and Hogarth. The defection would mean not only the loss of a conspicuous partner, one with clout and potential sales outlets for Sunrise, but $1.65 million as well, the utility's half of the study for manufacturing.

Finding ways to cut the costs of making lightweight composite bodies in large numbers, which the grant would pay for, was absolutely imperative if Sunrise was going to sell for a reasonable price. To date all the talk about 20,000 vehicles selling for $20,000 each was loose and speculative. No one had yet devised ways to make large, complex composite shapes in volume at costs competitive with steel, or even aluminum.

During the first months of 1995, things continued to deteriorate between Boston Edison and Solectria. Some of the secondary partners also had qualms about the whiz kids in Wilmington.

"James Worden should be courting us rather than pissing us off," said Dave Maass, business manager at Dow-United Technologies, one of the composites partners. Sunrise was intellectually challenging and economically promising, Maass added, but working out intellectual property issues and ownership questions with Solectria was exasperating. "I've lost faith," he said, shortly before Dow-UT cut its ties with the project. "The young management doesn't appreciate the magnitude of what they're trying to do."

That spring Boston Edison officially withdrew from the design for manufacturing study, citing worries about deregulation. Boston Edison did promise Solectria a vacant building and a seven-acre site in South Boston for manufacturing use, if the Wilmington crew could get Sunrise that far. BECO also remained committed to finishing the prototypes of phase one. Meanwhile, another composites partner, Draper Laboratories, also withdrew from the project after key personnel left the company and some issues between Draper Labs and Sheila Lynch at the NAVC couldn't be resolved.

During this time Worden, encouraged by Wayne Kirk and Mark Dockser, relented in his opposition to the NIST grant application. With Kirk, Dockser, and Ken Sghia-Hughes putting in a down-to-the-wire effort, the application went in on the letterhead of Lynch's NAVC.

The Dream Is in Trouble

But quite suddenly James Worden's dream of manufacturing Sunrise was in serious trouble. Solectria needed knowledge, it needed maturity, it might need one of the Big Boys as a partner, as MacCready said it would to survive in the car business.

The situation gave rise to an important question: If Solectria had grasped the magnitude of what it was trying to do, would the company have been trying to do it? If Worden had aggressively courted bigger companies to make manufacturing easier, would he have been the same inspiration to his employees? Probably not, at least during the years Solectria was growing and Sunrise was developing as a prototype capable of being manufactured. The conventional step of seeking out a bigger company earlier would have been logical for a conventional mind. But the mind in the head of the twentysomething engineer who dreamed of firing a revolution was not conventional, though it was being pushed powerfully in that direction.

At any rate, blame for Solectria's travails in early 1995 could not be assigned solely to Worden or to the immaturity of the company and its troubles with growth, even if it was appealing to do so—to write Solectria off as a cluster of whiz kids with attitude, as opportunists afraid of sharing control, of a research and development firm maxed out at fifty employees. Reality was much more complex, more kaleidoscopic. The regulatory merry-go-round anchored by zero-emission mandates refused to stop and clarify direction or intent. The continuing onslaught by auto and oil interests against the mandates, their insistence that clean-car technologies were not market ready, their arguments put forth by experts that improved gasoline engines were better for atmospheric health, global warming worries, and smog made small players in the green car game edgy and skeptical.

For Worden, however, staying in the game was not an elective. The engineer, the perfectionist, the cautious risk taker who didn't want to screw up Sunrise was ruled by necessity and by technological realities; he feared for a breakdown of his company and of his dream if timetables and technologies were pushed harder than either could bear. In this regard he agreed with the Big 3: Getting a vehicle to market before it was ready would be disastrous. Worden may have lacked gray hair, but he had plenty of gray matter. And he had iron-willed conviction, as well as youth, on his side. And youth fires revolution.

The loyalty of DARPA to the Sunrise dream was a godsend as well. "Without DARPA money," Worden said, "Sunrise would have stayed a dream, it would have stayed like the Flash."

The agency had a new administrator, John Gully, a civilian who had replaced Major Cope, who retired and went to work for U.S. Electricar, a big surprise for Solectria. John Gully liked Solectria, he said, but he didn't think Worden and his crew understood that little guys didn't build cars in America. Big guys did. "That's a hard reality for my young friends to accept." Another hard reality was this, Gully added: "The big guys, when it comes right down to it, will stomp the little guys if necessary."

One thing was certain. Little would change in the global auto industry or in the battle for a clean atmosphere because of Solectria alone. Making cars, dirty or otherwise, continued to be, as Paul MacCready said, a Big Boys' game, one refereed by regulators (at least in some countries) and played for the masses. In this global arena of Goliaths, the Davids certainly had their vision and commitment. But tremendous capital and little commitment still ruled. Any alliance that the smaller companies, the Davids, intended to forge to be able to compete had to have its act together, its capital firm, its vision unified. The alliance that had brought Sunrise so successfully to Anaheim unfortunately didn't fit this description. Now it was out of synch, and out-of-synch alliances don't lead much of anything.

Forces Behind Automotive Revolutions

Historically, two phases of America's automotive past could truthfully be called revolutionary. The first had been driven by changes in production, the second in marketing.

The first revolutionary phase occurred between 1896, when the Duryea brothers sold their first car, and 1941, when production was suspended and the American auto industry diverted its strength to winning World War II. Those fifty years spanned the era of the car as hero: its origins, adolescence, and wanton youthful decades. In those years auto making grew beyond anyone's imagination, became a bedrock of America's economy, and spawned heroes like Henry Ford, who introduced the Model T and the assembly line, Billy Durant, who consolidated small companies into General Motors, and Alfred Sloan, who transformed GM into "the archetype of the depersonalized, decentralized corporation run by an anonymous technostructure," as automotive historian James Flink wrote in *The Automobile Age.* The era rumbled forward primarily on the strength of industrial change, from how the parts of cars were made to how they were pieced together. Efficiency gains, vast factories that introduced the assembly line, and huge networks of interdependent suppliers made better cars in ever-greater numbers at ever-lower prices played key roles.

The American industry's second revolutionary era spanned roughly another half century, from 1946 into the early 1990s. This was the era of runaway horses. The revolution was in marketing, the focus not on production but on design. During the first decades of this era, cars grew in size, becoming more ponderous and chromed. Models proliferated. It was a time of technological inertia (engines, brakes, suspensions, and so on got bigger because of the size of most cars but changed little technologically), stylistic excess, energy indifference, and big profits. Between the mid-1960s and the early 1970s, the era ran head-on into the combined forces of regulation, which wrestled with safety, efficiency, and air pollution issues, and foreign competition, which intensified the pressure to change because other automakers, especially in Japan, were prepared for new global realities that their American counterparts were denying. This era was coming to an end in the mid-1990s. Another round of changing global realities dictated the dawning of a new era: the era of the green car, the ZEV.

The new era was one of limits, responsibilities, and industrial maturity. It was being forced to recognize, albeit with much resistance, the consequences of automobility. For status quo companies, though, the new era was hard to swallow. The new era had implications for the future of the world that its predecessors had lacked. The new era was no longer dominated by the American automakers but belonged to the world at large. A key question was this: Could the global auto industry lead away from, rather than toward, destruction of the atmosphere and of the life-support systems with the introduction of greener cars? Not that any car was truly green. Not when "green" was supposed to be synonymous with clean, with sustainability, with equality. An automobile, by its very nature, regardless of its fuel or recyclability or its mode of production, was the essence of personal power, of consumption. That truth was what had made the twentieth century the century of the car. If forced to choose, most people around the globe would probably give up their legs for wheels. Legs, like bearing witness with eyes, had lost power to wheels, as eyes had given up the power to shape events to television. Give someone a car with a TV in it: wouldn't he or she be in pretty good shape to engage the modern world? But greener, cleaner, lighter cars were a place to start. If their technologies could not reverse the deterioration of the atmosphere and global warming indicators, then the world really was in a catastrophe mode.

In the mid-1990s many places in the world were putting the question of how many cars their roads and air could handle to the test. But no country was doing it on the scale China had announced.

Until the early 1990s, a private person couldn't even own a car in China. In 1994, Mercedes 600 SELs (many of them stolen in Hong Kong and smuggled across the border) were dodging potholes in Beijing and Shanghai, and the Chinese government, an unpredictable and bureaucratic cadre of leaders who wanted to marry free-market forces and state planning, while downplaying a deep distrust of their own people, said they intended to jump-start an auto industry. China had a quarter of the world's population, approximately 1.2 billion people, and only small and scattered automakers with archaic machinery. Now China was going to be a world-class automaker, it said, relying on its classic five-year-plan approach.

Most of the world's Big Boy automakers wanted to give China a hand. Here was an untapped market they could sink their teeth into. China desperately needed their technologies, their expertise, and their money to make things happen. In 1994 and 1995, the Big Boys made connections, shipped over prototypes, courted bureaucrats, and consulted with experts on Chinese politics, manners, and business procedures. GM even sent over an Impact. "They thought it was from Mars," said the Impact's chief engineer, Jim Ellis.

James Hogarth, the eternal optimist, said he hoped the Chinese government recognized the flaws of an oil-based automotive industry. He wanted to see China leapfrog the ICE and go electric. He said, "Our hope is that China, India, or another one of those countries in Asia will commit to one, two, or three years of annual production of the Sunrise, so we'll have the capital leverage to build a manufacturing plant. Then we'll be in business."

Environmentalists, along with scientists and some politicians, were worried about China's automotive ambitions. Twenty, fifty, a hundred, or two hundred million more gasoline-burning cars—what would be the impact on air quality, oil supplies, and global climate change, not to mention on China's food supply as roads and parking lots displaced rice paddies in the great march forward? Projections made by the Chinese Academy of Science, an internal organization not given to harsh criticism of the country, said global climate change brought about by increasing carbon dioxide and other greenhouse gases over the next thirty to fifty years could flood Shanghai, Guangzhou, and low-lying coastal regions, displacing 67 million people. An independent worldwide organization, the Intergovernmental Panel on Climate Change, which consisted of 2,500 climate scientists, issued a warning about a coming catastrophe from carbon dioxide emissions, exacerbated by China's industrial great move forward. At the IPCC's international meeting in Berlin in the spring of 1995, however, foot-dragging by the United States and Japan, together with resistance by the oil-producing nations, curbed efforts to take a stronger stand on the impending crisis.

What crisis? Experts hired by big oil and big auto insisted that the seriousness of global warming was being blown out of proportion. The claims and counterclaims about the validity and threat of global warming seemed to be repeating the smaller but prophetic scientific standoff between Professor A. J. Haagen-Smit and the Stanford Research Institute in the early 1950s. Nowadays the truth about ozone was known, though still debated. The truth about global warming was just debated, although knowledge of its potential to alter life was rapidly accumulating. One difference was the scale of the two phenomena. The magnitude of the possible problems associated with global warming dwarfed those that had resulted, and continued to accumulate, from ground-level ozone increases. With global warming the experts paid by EXXON, Shell, ARCO, and other oil companies denied the evidence not of one but of thousands of scientists. Otherwise, the tactics used in the ongoing battle of words, charts, meetings, and the media were much the same, if more sophisticated, as in Haagen-Smit's day. The critics attacked with new evidence, high emotion, and usually limited resources; the defending experts denied the responsibility of their clients, called for more studies, defused the situation with disinformation, and said repeatedly that not enough was known to draw alarming conclusions or to take action.

Scientists' warnings and James Hogarth's hopes notwithstanding, the chances of thousands of EVs dodging potholes in Beijing, or electric rickshaws serpentining down the alleys of Shanghai, if the water didn't rise and short out the batteries, were slim. The people running the world's oil and coal companies knew that change could be minimized by the effective use of political lobbying and disinformation. They also knew that China's leaders had little patience with the concept of the earth having limits. A case in point: In late 1995, Lin Zongtang, vice chairman of the environmental protection committee of the National People's Congress, defended China's all-out rush toward industrialization piggybacked on its vast coal resources. "About 80 percent of the world's pollution is caused by developed countries," he said, "and they should be responsible for these problems." One idea floating around was that if the developed countries were so worried about China, they should pay to clean it up.

At any rate, imagining the scale of an improbable but intoxicating EV market in China was something that gave Sunrise a boost. It needed it.

A Phantom Car

In early 1995, Sunrise was a phantom car. It was illusory, a project whose reputation, buoyed up by Boston Edison's promotion at EVS-12 and the audacity of the marketing boast, far exceeded its reality. In Wilmington the one-off prototype was not road ready yet.

Gradually, though, the prototype was pieced together. In the cramped quarters, electrical and mechanical components went in. A suspension system was attached. The power train and electric motor were integrated into the car. Some systems worked smoothly, others didn't. "One of the big challenges was the doors," recalled electrical engineer John Rogers, who joined Solectria shortly after EVS-12. "The doors kept not closing right. They'd blow open in the wind and rip themselves off the vehicles." The solution to the problem was a stop rod designed by an engineer. "If you notice decals along the seam of the door in prototype one," Rogers said, "those are actually covering damaged portions of the door."

There were other problems to solve as well: the car needed an interstate to make a U-turn; the plastic side and rear windows kept popping out in the sunshine because of expansion; the accelerator pedal stuck at times and at other times caused the car to lurch forward drunkenly. The problems were solved primarily by the team, whose membership had changed considerably. Stylist James Kuo had left to work on a real phantom car, the Batmobile, for Warner Brothers in Los Angeles. Engineer Ed Wogulis had quit. Modeler Dave Blair and project engineer Wayne Kirk remained, working often with John Rogers, whose principal job was to marry the composite body to the electronics systems, and Ken Sghia-Hughes, now in charge of safety and other regulatory requirements.

Like most Solectria employees, Sghia-Hughes, who had volunteered his Geo Metro as Solectria's first conversion back in 1990, wore many hats. He was "the regulation guy" and the safety guy. Safety was a big deal. Few things could torpedo a company in the automotive business faster than a bad safety record. Composites multiplied the safety issues considerably because they reacted differently in an accident than did either steel or aluminum. For instance, in a crash of a composite vehicle, practically all the kinetic energy would be transferred instantly to the occupants, unless safety restraints and interior design were tailored to the realities of the stiffness of composites, or crumple zones were engineered into the front end of the car, as they were in steel cars to reduce the kinetic energy transfer. During Sunrise's development, Sghia-Hughes discovered that there were few data banks with information about composite vehicles and that most regulations had been written with steel cars in mind.

Under Sghia-Hughes's direction, and with the cooperation of the University of Lowell, which had a crash test facility, the composite laminates in Sunrise were designed and redesigned to handle energy in a crash. Special front ends, called "noses," were fabricated and tested. Eventually, Sunrise prototype number two, which was built for crash testing, hit a wall at thirty miles per hour and passed the federal safety standards for frontal impacts.

Months before that, however, the prototype P1 had debuted as a roller. This happened at the spring meeting of the DARPA consortia members. In contrast to the big splash in Anaheim six months earlier, which had been made possible, according to James Hogarth, by the fact that Boston Edison spent $150,000 on press and media hoopla, the unveiling of the prototype was pretty restrained and low-key. DARPA held the meeting in Washington in the hopes that consortia members would lobby their congressmen about advanced technology appropriations. The new Congress was zealously going after pork in the federal budget, and some congressmen saw money spent on EVs and their technologies as pork. At the Grand Hyatt Hotel, where the DARPA sessions were held, anxiety over future funding was high. Lynn Jacquez, a lawyer who represented CALSTART, the California consortium, said that the political mood in Washington could be compared to a flavor: "garlic with a twist of lemon." This year's appropriation to DARPA had been salvaged, she said, but next year's would have to be fought for. She reminded her audience that the real business of Congress was money, and she urged everyone in the room to visit and lobby their congressmen. "Presume total ignorance about electric vehicles," she said. "Give them the ABCs. Don't lapse into the jargon of the trade. Eyes will glass over."

On the final day of the sessions, a sticky, hazy, blue-sky day, consortia vehicles went on display on the Mall, with Sunrise the star of the ride-and-drive offerings. As the cars gave the curious quick spins around the few blocks of trees and grass, joining the scene were a few midranking bureaucrats from the Departments of Defense, Energy, and Commerce. They wandered around, checked out the display cars, perused the technology exhibits, and talked nervously about pending cuts in the clean-car pie and how they might directly affect them. When they ducked into a large white tent that contained fuel cell and flywheel and battery exhibits, they passed Catherine Anderson.

The former MIT dream team mechanic who had rescued James Worden and kept his cars running now worked at AeroVironment with Paul MacCready. Anderson was in Washington because she was AeroVironment's program manager for something called the Joint Tactical Electrical Vehicle, or JTEV. The Jeep-like yellow-and-black prototype with a tubular roll cage to protect the driver and passengers in case of a rollover was parked by the front of the white tent. Anderson explained how it could climb a 60 degree slope and had an electric drive mode that would allow it to sneak up on the enemy silently and undetected. Or so the U.S. Marine Corps hoped, for it was partially funding the JTEV's development program. Working with the marines was a bit different, Anderson said. "My sex and age threw the uniformed guys at first, but my expertise soon leveled the field."

Parked on the other side of the entrance to the white tent was the AFS2000, a sleek, flywheel-powered four-door sedan prototype built by American FlyWheel Systems, Inc. A flywheel is a mechanical battery that works according to Newton's first law of motion: Things in motion tend to stay in motion. Basically, a flywheel stores kinetic energy in a spinning mass, then converts the energy into electrical current on demand.

Flywheels, along with fuel cells, were promising alternatives to batteries as the source of energy for electric cars, although size, reliability, and cost problems remained formidable. Fuel cells had the lead temporarily. Recent advances in fuel-cell technologies had made the units, once large and expensive, small enough so that Daimler-Benz AG, the parent company of Mercedes-Benz, would soon unveil the world's first fuel-cell-powered car in May 1996. Fuel cells remained expensive, however, partly because of platinum, a rare metal they used as a catalyst to speed up electrochemical reactions, and polymer membranes, which act as electrolytes through which current flows.

A fuel cell worked in a simple and quite beautiful way. Instead of its fuel burning and driving moving parts, as in a gasoline engine, a fuel cell converted a fuel's energy into electricity—with no moving parts. By means of an elegant electrochemical process, a fuel cell peeled electrons off hydrogen atoms. The electrons veered away as an electric current, which could be used just like the electric current derived from batteries. The hydrogen passed from an anode, with a negative charge, through an electrolyte, to a cathode, with a positive charge, and then reunited with oxygen to make water. A fuel cell was thus virtually emission-free. It ran at low temperatures. "Fuel cells sound almost too good to be true," said David Swan, assistant director of the Institute of Transportation Studies at the University of California–Davis.

As with a gas car, refueling a fuel cell could be fast, taking only a couple of minutes. And fuel storage challenges seemed reasonable. A thirty-five- to forty-gallon aluminum canister, weighing 180 pounds, could hold enough hydrogen to give a fuel cell car a 250-mile range.

Still, in 1995 a practical fuel cell car was far off in the future. And flywheel developments lagged even further behind, the AFS2000 notwithstanding. That the car could be made and sold for around $30,000, as the manufacture boasted, was unlikely.

All afternoon, as the public mingled with the alternative-vehicle professionals and the Washington bureaucrats, ride-and-drives continued. In addition to Sunrise, the curious got introduced to a new generation of the Swiss-made Horlacher AG pickup, a Suntera Solar Chariot from Hawaii, and a fuel-cell-powered bus. No elected officials—no senator, no representative, no Bill Clinton, no Al Gore—showed up on the Mall to take a spin in a car of tomorrow, though. That was a bit troubling to the green car crowd, whether they were associated with DARPA or with the Partnership for a New Generation of Vehicles (PNGV).

James Merritt, the Department of Energy program manager for transportation technologies, had one plausible explanation for the poor turnout of politicians. Merritt said that most of the congressmen who knew much about alternative vehicles, and were supportive of their potential, had either retired or been defeated in the previous November's election. "The new guys have a lot to learn in a short period of time," he said.

Merritt's boss, an affable fellow named Bill Siegel, added that their department was handicapped. "We can't lobby!"

Presently, the Big 3 did it for them. The Big Boys lobbied for a federal bureaucracy because funding for the PNGV potentially benefited both of them. The tactic also caused problems, Siegel added. Congress was leery of the Big 3. The auto industry's history of self-interest and of least feasible retreat had come back to haunt it. "This is a difficult game," Siegel conceded. "To get these new vehicles we see here in Washington into production is going to take a revolution on the part of the industry."

The PNGV seemed unlikely to lead such a revolution given the mood of Congress, which was on an antitechnology crusade, or at least on an anti-government-involvement-in-new-technology crusade. Arguments for continuing the big business/big government alliance said it had already been initiated and the partnership was not a mandate. Arguments opposed said the PNGV was too big, partisan, and something private industry, without the cooperation of the Departments of Energy and Commerce, could do faster and cheaper on its own.

James Worden agreed. The Department of Energy was a poor choice for leading change, he said, unless you wanted it to go at a snail's pace. "The DOE spends huge sums from a distance," he scoffed. The Pentagon's DARPA was much faster and more efficient. "DARPA gets right down into the ditches and takes risks," Worden said.

Getting into the ditches and taking risks had its perils, however, especially if your projects succeeded. In the bureaucratic world, too much success could become an embarrassment. Too much success could kill a project faster than plodding progress.

To date DARPA had been enjoying some success with its electric and electric-hybrid vehicle projects despite their relatively low budgets. In doing so, the Pentagon agency had stepped on a few toes. Now, with the bureaucratic PNGV hungry for a lion's share of shrinking new technology funding, DARPA had to be crafty and cautious if it was not to be eliminated from alternative-vehicle development work. If DARPA moved too fast, or duplicated work the PNGV was doing or could do, albeit slower, or if the military agency failed to clarify the dual-use applications of its projects, Congress might cut its funding, and the funding of its consortia, with one quick slice of the budget knife.

Given that possibility, DARPA administrator John Gully judiciously praised the Department of Energy's role as a key player in clean-car technologies. In contrast, Gully said, the little guys DARPA worked with were fast, lean, hungry, and smart. But could they make cars in America? He thought not. Making cars for the public, with its economies of scale, "is an art," Gully declared. You couldn't compare it to defense work, where only a few of anything had to be built at a time. No doubt about it, he continued, the Department of Energy is a huge bureaucracy, but it sometimes takes bureaucracies to get things done on a large scale. "These big bureaucracies are rather inefficient, but if we put them with the Big 3, they can sell cars. We don't know if James Worden and Solectria can do that."

Despite his doubts that Solectria could make cars, Gully liked the company as a Gen-X powerhouse research and development operation the Pentagon could plug into to push advanced systems at low cost. Because DARPA encouraged and financed such small companies, the Big 3 and other groups often got upset with the defense agency. "But," Gully said with a smile, "we're supposed to be out on the pointy end of the spear."

A Car Company Needs a Car

In May 1995, Solectria found itself on the pointy end of a spear of its own. Sunrise P1 was a drivable prototype, but Boston Edison had pulled out of the design for manufacturing study, so the study's future was in jeopardy. The NIST grant that had been applied for with the Northeast Alternative Vehicle Consortium held out some promise to push composite processes closer to manufacturing. But the NIST grant was a long shot.

Basically, Solectria was a car company without its own car. It was a high-tech research and development outfit with a line of conversions retrofitted with cutting-edge components it designed and installed. But to grow beyond that, the company needed a product base. Converting Geo Metros and Chevy S-10s into Forces and E-10s was not a viable product base. The conversion business was good—the best it had been since Solectria got into it, actually—but it was an end game. All the Big Boys had to do was start building EVs seriously. That could happen soon. As Paul MacCready said, lots was going on out of sight with the Big Boys. Once the Big Boys committed to manufacturing, which GM and Peugeot were poised to do, with Honda, Toyota, Mercedes, and others in the wings, if Solectria didn't have a car of its own, that could be the end of the game.

Could the company survive as a supplier to big companies? Could Solectria be sold?

For the short term such questions could be avoided because of the strong sales of conversions. Solectria had U.S. Electricar to thank for that stroke of good luck. The California-based company, Solectria's main competitor in the conversion business, had overextended itself and could not fill its orders. Never a technology leader, U.S. Electricar had been more of a systems integrator. But without cutting-edge technology, the company had floundered. Major Richard Cope, having retired from the marines and placed DARPA in the hands of John Gully, had been brought in to halt the slide, but to no avail. "Major Cope thought he could step in there and be a savior," Worden said. "But this is a tricky business. At Solectria, we're kind of scared at times that the people with the cheesiest cheap product are getting an edge on us. But it always comes back to our technology; it really is our only and main strength."

In Wilmington, Solectria continued to vigorously work on three elements of its strength: a new universal motor controller, an integrated motor gearbox package, and a truck axle system. All would soon reach the market in conversions.

But what Solectria truly needed wasn't improved conversions. It needed its own car.

Worden knew this. He knew that, as in racing, if you crossed the finish line first in getting a product to market, the media gave you the checkered flag, positioning you in the public's mind as a winner. If Sunrise rolled out of a small assembly plant as an affordable, crashworthy, reasonably performing sedan before the Big Boys rolled out their vehicles, the marketing value would be tremendous. On the other hand, now that Solectria was venturing closer to the realities of manufacturing, things were getting a little scary. Manufacturing required new skills, additional players with experience, capital, construction capabilities, and distribution.

In a way Solectria was stalling. Maybe the learning curve had become too steep, the leaps too broad, the financial abyss too deep. After all, it was only ten years since James Worden had been a senior in Arlington High School, bothering local hardware store owners for throwaway sprockets and gears, and building cars in his parents' garage. Now he was the head of what was arguably the most innovative small car company in America. But if Solectria was to grow, he had to make big leaps, have tremendous luck, and still sustain his vision. Even then he might become only a minor footnote in the history of the American auto industry.

Presently, on the average, Solectria completed one conversion a day. Just one. If the company was to make 20,000 Sunrises a year, it would have to build 80 a day. How was such a leap going to happen? Even making 5,000 Sunrises a year, or 20 per day, was a vast undertaking. To do that, some of the strengths and habits—seat-of-the-pants victories, loose schedules, an insistence on control, no debt—that had gotten Worden and Solectria to a place where manufacturing was even a possibility had to be left behind. In the terminology of modern science, James Worden was encountering "incongruity"—the end of the known road, an abrupt terminus, a place where he had to jump. "And incongruity, when it changes the way a scientist [substitute engineer] sees, makes possible the most important advances," wrote James Gleick in *Chaos: Making a New Science*. Incongruity was a place that was so dangerous and threatening that it had the power, for a few, to change the way they saw things and demanded from them not incremental improvements, such as the auto industry cherished and defended, but "the exceptional, unorthodox work that creates revolutions," wrote Gleick.

This place from which Worden had to jump, if Solectria was to make cars, was intimidating. But he had to go for it. The young engineer who still, as much as anything, loved to simply race cars that were fast and clean and efficient, had to leap.

17

May 1995: The Seventh NESEA American Tour de Sol

C. Michael Lewis

The original Sunrise prototype winning the commuter category of the Seventh American Tour de Sol, in Portland, Maine, May 1995.

Why aren't they out here cheering?
—James Worden, passing a vacant Ford dealership during the Seventh Tour de Sol

Worden raced Sunrise for the first time in late May, in the Seventh American Tour de Sol. The tour began in down-and-out Waterbury, Connecticut, onetime brass capital of the United States, and ended in the trendy seaside town of Portland, Maine. Entered in the commuter category, against twenty other competitors, Sunrise ran on Ovonics nickel-metal hydride batteries, similar to those used in the Solectria Force RS the year before. The car was favored to win its category, but on the second day of the tour, between Northampton, Massachusetts, and Keene, New Hampshire, Worden got ticketed for speeding. Not by a cop, but by a race official. Penalized, Sunrise dropped into second place.

The next day, racing southeast of Keene, Worden was impatient; he wanted to regain first place. He was tailgating the pace car, a '93 Solectria Force owned by Virginia Power and driven by Solectria head truck engineer, Andy Heafitz. Worden kept lifting his right hand from Sunrise's steering wheel and flashing a signal—five fingers, then a zero formed by his thumb and forefinger. The low-tech gesture meant: Go fifty miles per hour, Andy. You're going too damn slow!

Heafitz ignored his boss. He'd caught enough grief the day before about speeding in the pace car.

In Sunrise, Worden grumbled about the slow speed to his navigator, Ritu Monga. Monga, eighteen, was from New Delhi. A senior at the Ethel Walker School in Connecticut, she was on the school's EV racing team, which was piloting a Solectria Force farther back in the pack. In Monga's lap was a clipboard that held data sheets, directions, and maps.

Monga advised Worden where to turn in Jaffrey, in Rindge, in Ashby. Visually, the route was postcard New England. The vehicles, which this year included more hybrids and a recumbent bicycle pedaled by a longhaired marathon biker named Me, traveled narrow, bumpy roads, passed through low hills, circled commons, went by white steeples that cast pointed shadows. As he approached the Massachusetts border, it occurred to Worden that he just might be driving the most efficient four-passenger sedan in America. With the Range-Power Control on "economy," the 1,600-pound composite car was getting the equivalent, energy-wise, of seventy miles per gallon. The performance put the prototype, which had cost less than $1 million to date, where the Partnership for a New Generation of Vehicles aimed to be in 2003, after spending $1 billion—if Congress continued the funding.

That was anything but certain. Congress had eliminated the Office of Technology Assessment, which reported on how funded technologies were doing; the office was no longer needed if the government wasn't funding cutting-edge work. Congress had also cut the Department of Energy's advanced-transportation technologies budget by 30 percent for fiscal year 1996. And given the reluctance of a majority of the members to act on reducing America's carbon dioxide emissions to set an example for slowing global warming trends, climate change was nothing Congress was very worried about either. Meanwhile, oil imports continued to increase, with the Persian Gulf acting as the world's main oil bank. As imports rose, America's trade deficit rose with them. Finally, with so much cash going to Saudi Arabia for oil, the need to prop up that unstable, oppressive monarchy would likely grow stronger, requiring more military aid and international mischief in the name of democracy and progress. Such spending seemed acceptable to Congress, with its antiregulatory bent. It was wise to delve into foreign affairs in a way that controlled outcomes; just leave domestic problems alone and let business handle them.

It was strange to think that Sunrise, this one-off prototype rolling silently and emission-free across the Massachusetts border, held so much promise to counter negative forces if it could make it into the streets in significant numbers. Or if vehicles like Sunrise could—and make it relatively quickly, since the momentum behind air pollution, global warming, and oil depletion was so powerful. In a way, Sunrise could be seen as a bowsprit, as a steering device, on the prow of an oil tanker. The oil tanker was the way America, and the world, did transportation. Veering from that path would take a while. It would demand the tortoise-slow workings of the huge bureaucracies, the millions of dollars made available by maverick defense agencies, and ultimately the machinations of the Big Boys. For after all, even Solectria unbound, Solectria being a Microsoft-like success, wouldn't satiate the world's hunger for wheels. Being on the prow of the future's momentum, though—what an exciting, dizzying place to be.

Dizzying, but in this race, noisy as well. Sunrise's sound-deadening foam, thirty pounds worth, had been removed for greater efficiency. You could hear squeaks from the suspension, the low hum of the motor. Road hum from low-resistance tires, along with the sounds of traffic, penetrated the composite skin. Inside that skin the interior was austere but comfortable, a blend of gray and pink. The air-conditioning was off, for efficiency reasons, and the windows were up because they were screwed in place. The screws, backfilled and covered with electrical tape, had been the solution to the problem of the windows popping out in the heat because of differential expansion.

Suddenly, a large Ford dealership appeared ahead. Rows of windshields and headlights faced the highway. Passing them, Worden shook his head. "Why aren't they out here cheering?" he asked rhetorically.

He knew why. No one at Ford had told the dealer to get the salespeople and mechanics outside to cheer, to hang a banner or two, to climb in all those ICEs and lay on the horns. Maybe it would have happened if the ZEV mandate was dead. But it wasn't—not yet.

A moment later, a Ford Ecostar passed the ghostly silent dealership.

Today's leg of the Tour de Sol ended at Minuteman Science and Technology High School in Lexington, Massachusetts. Worden drove beneath the finish line banner behind Andy Heafitz. "I've only used one-third of the batteries," Worden told Monga. "For seventy miles! Isn't that incredible?"

Ritu Monga nodded in agreement.

Never far from his own sense of awe over the magic of electricity, James Worden grinned. He wheeled Sunrise into the display area. Some high school kids stood nearby, furtively eyeing the yellow-and-brown car. Worden got out. He was wearing a Solectria T-shirt, shorts, and sneakers. He stretched his back. Soon Solectria's most sought-after media person was talking into microphones, facing cameras.

A Healing Touch

His white shirt starched, his tie tight, Don Walsh, Boston Edison's economic development director, stood in the parking lot at the Minuteman Science and Technology High School. Walsh was a visible symbol that Boston Edison harbored no hard feelings toward Solectria. With Walsh was Dave Dilts, who was being eased in as head of the utility's EV program while James Hogarth was being eased out. Though Boston Edison was no longer a partner with Solectria in phase two of Sunrise, the design for manufacturing study, it remained a partner, with DARPA and the Northeast Alternative Vehicle Consortium, for phase one, the completion of three prototypes. Boston Edison also retained distribution rights for the car, if it was made.

Dilts wondered if that could happen. Like Hogarth, he had his doubts about Solectria's management. "What that company needs is a prick in the back room," he said on occasion, "a real-world manager, somebody who can make people perform. Everybody at Solectria is a nice guy."

Or almost.

About a half hour after Sunrise had wheeled into the high school parking lot, Walsh came strolling over to Dilts, who was manning the Solectria/Boston Edison display table with Sunrise flyers stacked on it and weighed down by stones to keep them from blowing away in the May breeze. Sounding not very pleased, Walsh said in a low voice, "James wouldn't let me sit in the car."

The diplomatic Dilts assured the slightly perturbed Walsh that it was nothing personal. "If James let you sit in the car, he'd have to let all those high school kids sit in it too."

Walsh took another look. A couple of kids were peeking in the windows, off which sun glared. Walsh didn't seem to totally buy Dilt's explanation, but he smiled at a passerby and let the incident ride.

Walsh left Dilts at the display table and wandered around the hot, busy parking lot. His job was to facilitate healing, not to get irritated at Worden, like Hogarth had. Hogarth still wanted to play hardball with the whiz kids. Walsh didn't share Hogarth's animosity. Spreading positive energy, he worked the crowd some more, talked to high school kids, adults, and racing team members by their vehicles now on display after completing the leg. Walsh said things like:

"Making electric cars has been a dream; Solectria is making it a reality."

"There's a nice marriage between EV manufacturing and economic development."

"EV manufacturing is going to happen."

But how?

At this point, the ZEV mandates in California, Massachusetts, and New York seemed doomed; continued DARPA funding for EV development remained shaky; and the Clean Air Act Amendments of 1990 continued to come under attack in Congress. Boston Edison itself was staying on the sidelines and watching Solectria struggle with manufacturing issues. If Solectria could demonstrate some business maturity, if it could work out the glitches with the prototypes, then maybe Boston Edison would again play ball with the company. Until then it could use deregulation as a foil, claiming, not altogether untruthfully, that deregulation was turning its world upside down. To keep the relationship going with Solectria, the utility continued to dangle a 40,000-square-foot factory facility and a nearby seven-acre lot, both in South Boston, as a possible future manufacturing site.

The next day, Sunrise, with Worden at the wheel, set a new distance record. Doing extra laps in Dover, New Hampshire, Worden held a roll of quarters in one hand, tossing them into the toll booth, which he rolled through slowly, not coming to a complete stop, capturing that extra little bit of efficiency. He finished the day with 238 miles on a single charge. And Sunrise was back in first place in the commuter category.

On the final day of the tour, while the vehicles were still on the road between Dover, New Hampshire, and Portland, representatives of the Maine branch of the American Lung Association handed out report cards on the state's air quality. Maine got a D. The reason for the low grade? Ground-level ozone, or smog. Daily, the ozone transport phenomenon dumped tons of smog on this state. Industry and tens of millions of vehicles in downwind states, from Massachusetts to as far away as Virginia, were largely to blame. The smog they produced moved by both land and sea (smog travels especially well over water, riding the prevailing breezes), and it gathered over Maine's sandy beaches and rocky coast, as well as over larger cites like Portland. The phenomenon evoked concern in a state with a vested interest in air that neither upped the cancer count associated with languorous summer vacations on sandy beaches nor harmed forests, which were the source of hundreds of wood products.

Recently, Maine had tried to lower ground-level ozone with an inspection and maintenance approach. Together with cleaner cars and cleaner fuels, inspection and maintenance was a key strategy of the federal Clean Air Act Amendments of 1990 to get air pollution down. It hadn't worked. Poorer folks in older cars got hit with high bills for pollution controls, while owners of newer cars typically drove off free. This "fiasco," as Maine congressman James Longley called it, had moved Governor Angus King to suspend inspection and maintenance. This in turn left service station owners holding expensive testing equipment, Maine in violation of the Clean Air Act Amendments, and air quality unchanged. With just about everyone involved irritated, the U.S. Congressional Subcommittee on National Economic Growth, Natural Resources, and Regulatory Affairs scheduled a meeting to air out matters at Southern Maine Technical College in South Portland on, coincidentally, the same morning the Tour de Sol finished in Monument Square downtown. James B. Longley (R-ME), David M. McIntosh (R-IN), and Collin C. Peterson (D-MN) sat in chairs on the basketball court at the technical college, listening to testimony from environmentalists, antienvironmentalists, joggers, gas station owners, and others. At issue, in McIntosh's words, was whether Maine had been used as a "guinea pig" for clean air regulations. The guinea pig allusion referred to the fact that Maine was supposed to be acting as a model, a place where clean air regulations such as inspection and maintenance were tried first to validate the Clean Air Act Amendments of 1990.

Maine, located at the "tailpipe end" of the Northeast's air woes and on the opposite side of the country from California, had supported the clean air movement in America for decades. Senator Edmund Muskie had championed the Clean Air Act of 1970, the one that had given the original 1963 federal standards some muscle and forced the auto industry to respond. Later, Senator George Mitchell had pushed equally hard for adoption of the 1990 amendments. The level of knowledge about the science and health effects of ground-level ozone shown at the meeting with Maine's Longley, Indiana's McIntosh, and Minnesota's Peterson probably would have appalled both Muskie and Mitchell. In response to one question, Collin Peterson, the ranking minority member of the subcommittee, told the audience he didn't know the difference between types of ozone, or even what made up ozone. At any rate, testifying at the forum were representatives for groups that wanted Maine to be a model for mitigating ozone and representatives for groups that felt Maine could not do much about the problem by itself. Former Olympian Joan Benoit Samuelson, a gold medalist in the 1984 marathon in Sarajevo, told the congressmen that on high ozone days she didn't even lace on her sneakers because the air was so bad. Complaints were voiced about the health effects of reformulated gasoline, now being sold in Maine, and about the fact that Maine, downwind of the industrialized Northeast, couldn't do much about ozone transported often from hundreds of miles away. All in all, the forum seemed to validate the fact that clean air, and how to keep it clean, was a complex, confusing issue that Maine was ill equipped to handle alone, especially given its geographic position in the Northeast. The testimony brought to mind the comment Governor King had made when he suspended the inspection and maintenance program: "If we're ever going to solve this clean air problem, it has to be a national solution."

He was right. On the other hand, if states refused to act on their own, would a number of them do much collectively? To date, collective action had not happened to any significant degree unless mandated by the federal government. In the gym in South Portland, the three congressional members of the subcommittee investigating regulations didn't do much to support either state or federal action. They did fan fears of regulatory ineptitude, nodded in sympathy with those complaining about the financial unfairness of the inspection and maintenance program, and thanked folks for coming.

Several miles away, in Monument Square, machines that could lower ozone and carbon monoxide and carbon dioxide emissions, as well as reduce dependency on foreign oil, were driving through a colorful balloon arch, one after the other. Though no congressmen were there to greet the vehicles, plenty of kids were. Recognizing the best of these cars, awards went to Sunrise in the commuter category for best sedan using advanced energy storage; to two Solectria Forces in production categories; to an electric-propane Chevy S-10 pickup from the Mount Everett High School in Sheffield, Massachusetts, for best hybrid, and to the Ford Ecostar for best utility vehicle with advanced batteries.

18

Crippling the Clean-Car Mandate

Dennis Renault

In December 1995, bowing to political pressures and to the will of the auto and oil industries, the California Air Resources Board recommended that zero-emission vehicles not be mandated in 1998, as scheduled, but be introduced voluntarily between 1998 and 2003.

Agency behavior is partly conditioned by manipulative tactics of regulated parties
—Howard Latin, *Environmental Law*

The rest of that summer and fall things went from bad to worse for the alternative EV community. "The anti-ZEV forces smelled blood in the water," said Veronica Kun, a senior scientist at the National Resources Defense Council. "They were emboldened by the political sea change."

In California, ads, press releases, and studies continued to question the benefits of EVs. A clever, insidious strategy, greased by the Western States Petroleum Association, effectively stalled plans for EV-charging stations. So-called grassroots groups, which critics dubbed "Astroturf groups," consisting mostly of senior citizens worried about higher electricity bills, were organized and their fears exploited. The oil lobby facilitated protests, in which the seniors were bused to utility company headquarters and supplied with stationery on which they could write angry letters about possible rate increases to pay for EV-charging stations. Accused of seeding a bogus grassroots campaign with senior citizens, Douglas Henderson, executive director of the Western States Petroleum Association, said, "We're not ashamed of it at all."

In Washington, the EPA took a final, middle-of-the-road position on the Ozone Transport Commission petition. The agency ruled that it would do nothing—neither encourage nor endorse ZEVs over gasoline ones. If any state in the Ozone Transport Zone wanted them, it could pass its own ZEV mandate, as New York and Massachusetts had done.

Sheila Lynch threw up her hands in frustration. "For the past two years I've gone to ozone transport roundtables and panel hearings," she said. "I've been involved in lawsuits and state filings. The last OTC meeting I went to was last December. It ended nowhere. No consensus, no final meeting, no report, no anything. It was, 'We've stopped; you're all dumped.'"

After that, the agency made a decision not to make a decision—the outcome some had expected all along. Now the responsibility for ozone reduction in the Northeast, where Maine's experience showed that a single state was unlikely to fight the problem, was back on individual states.

"The EPA doesn't have any guts," said Trudy Coxe, Massachusetts's secretary of environmental affairs.

Two years earlier, it had looked like auto and oil interests might be in retreat, that clean cars from new players and ZEV regulations would be sustained. It had not happened. Using tried-and-true tactics—least feasible retreat, hard lobbying, clever disinformation in an inexorable tidal wave of persuasion applied by public relations, political, and scientific machinery well lubricated with money—the Big Boys had turned the contest into a rout. It didn't hurt at all that in Washington and across the Republic, conservatives had returned to power. Conservatives, even if they cared about EVs, wanted the marketplace to pull them into showrooms, not the government to push them there.

It was Holdfast Redux: the momentum toward a green car future was again in the grip of twentieth-century power brokers who didn't want to let go and face twenty-first-century complexities that were getting harder to deny: fossil fuel limits, global climate change, more ozone and carbon monoxide and health problems caused by air pollution, increasing military costs to protect oil kingdoms in the Persian Gulf and ensure a steady flow to America.

In Massachusetts, Solectria was getting bounced around but was staying afloat with its conversion business and with DARPA's investment in Sunrise. Even Governor William Weld, a strong ZEV mandate supporter until he decided he might take a shot at the presidency, was vacillating. Worden's advice for Weld was, "Hey, Bill, if you're not running for the presidency, think selfishly here; we can easily provide all the cars for your state."

Maybe. But either way, Weld was thinking about running for the presidency during the summer of '95. And so he was courting the Big Boys, just in case.

Potential delays were put in front of the ZEV mandate in Massachusetts. One was an announcement of performance "triggers"—things like a range of 150 miles on a single charge, zero- to sixty-mile-per-hour acceleration in fifteen seconds, and a price close to that of an equivalent ICE vehicle. If any automaker could show that it could meet the triggers, the ZEV mandate would remain as it was; otherwise, it would probably have to be changed.

On closer inspection, the triggers were a throwback to the 1960s, the days of technology following. Manufacturers were being asked to set the pace, to show they had the technology at an affordable price in order to justify government action.

The Big 3 simply ignored the Massachusetts triggers.

"The people in Trudy Coxe's office who came up with those triggers thought they were doing us a favor," Anita Rajan said. "They weren't. That we have to prove EVs can achieve those standards and then be sold for reasonable prices—I don't want to call it a burden, but it definitely puts a damper on the progress we would rather be making, which is on improving efficiency."

The triggers upset David Cohen, a state representative who had sponsored the Massachusetts ZEV mandate bill. That Governor Weld had come up with them without consulting the legislature was "absolutely outrageous," Cohen said.

Trudy Coxe disagreed. Defending her boss, Coxe adopted the Orwellian logic that the California Air Resources Board had adopted lately to explain why hybrids could comply with the ZEV mandate there. Whereas CARB spokesmen had consistently said, "Zero means zero," now they said, "Zero means almost zero." Coxe said, "The ZEV mandate in Massachusetts is inviolable . . . but it may have to be delayed."

The triggers and Orwellian rhetoric didn't stop progress at Solectria, but they were a burden. It was as if an unspoken challenge had been tossed out: okay, Solectria, show the Big 3 you can make this 150-mile-range, zero- to sixty-mile-per-hour, reasonably priced EV. The triggers, which were eventually abandoned, created more unneeded friction along the already rough road to manufacturing.

About the same time the trigger issue was hot, a confidential memo circulated by the American Automobile Manufacturers Association said that the Big 3 knew an EV market was shaping up, especially in California. According to the memo, there was evidence of a distinct shift from tentative support for EVs in 1993 to the perception in 1995 that they were becoming more practical and would work. The perception needed to be reversed if the mandate in California was going to be killed.

Exactly why was a little confusing. For in a Buick plant in Lansing, GM was experimenting with production concepts for the Impact. Sean McNamara insisted this was an experiment. GM was using a small corner of an old factory to see if the common Joe who built Buicks could build Impacts.

And if he could?

"We're looking at this as a business," McNamara said. "We'll take the pressure and do it right. Substandard production will give the market a black eye."

PrEView Impact was continuing eastward across the United States. The number of test drivers who didn't want to return the "oh wow!" two-seater with the modest futuristic design steadily increased. Once 300,000 miles were on the prototypes and 500 drivers had given their impressions, 85 percent said the car met their needs and had adequate range with lead-acid batteries. Ninety-four percent of the utilities involved in PrEView reported being satisfied with the program. Bob Purcell, GM's program manager for EVs, presented the figures at the North American EV and Infrastructure Conference in Atlanta in early December. Purcell knocked the ZEV mandate, said GM had spent $500 million on EVs and had 250 people working on them, but added that technical readiness and commercial willingness had to mesh before the Impact would go to market.

Following Purcell to the podium, Noel Bureau, deputy director of research for Peugeot-Citroen, said, "We are talking too much about EVs. We are doing it now!"

Maybe it had something to do with the fact the French are more philosophical. After all, it was Jean-Paul Sartre who said hell was not other people but other people in other cars in front of you in a traffic jam. And Roland Barthes, another French philosopher, had compared the modern car to a medieval Gothic cathedral, insisting both were "purely magical objects." Now it was the French strategy, outlined the year before in Anaheim at EVS-12, that launched the first EV in the public marketplace in the 1990s because somebody had to do it. The previous November, with price incentives to make costs competitive with similar ICEs, Peugeot 106s and Citroen AXs had gone on sale.

So the French message was the same as it had been in Anaheim: a team approach, involving automakers, utilities, and the government, was the way to introduce clean cars into the market. The message from the Big 3 was the same: Government should help but should let the marketplace rule. But the message from the little guys who had been plentiful in Anaheim was no message at all. In Atlanta the little guys had virtually disappeared. Having made the splash the year before in Anaheim, Sunrise was nowhere to be seen. James Worden didn't show up either. Only one representative from Solectria, vice president of sales, Mark Kopec, was at the symposium.

A Brief Comparison with the Mid-1960s

At this juncture, Solectria's story was worth comparing with those of the small, independent manufacturers of emission controls in California in the mid-1960s, the manufacturers who made the first add-on devices, only to have Detroit make a "sudden" breakthrough just weeks later. The breakthroughs put the independents out of the emission-control business, and put the business in the hands of those who had denied it was feasible. Thirty years later, the risks taken by independents like Solectria and Unique Mobility, the Colorado-based drivetrain builder that had come to the symposium in Atlanta, and a few others that continued to survive were more complex, extreme, and potent. What was similar about the situations was the fact that in both eras the independents pushed new technologies harder in order to comply with regulations, to give heart to those who had brought the regulations into being, and then were forsaken. By folding to enormous status quo–sustaining pressures, regulators in the 1960s and 1990s may have only been acting human. But they also demobilized the forces of change, leaving small cutting-edge companies with little to cut but their own futures—unless they could manage to befriend one of the Big Boys.

That had not happened in the mid-1960s. None of the independents went on to supply big automakers. In the mid-1990s, some independents were faring better. As already mentioned, AeroVironment had made the crossover. Unique Mobility had also cut some good alliances, most notably with Italian and Indian manufacturers of scooters and three-wheelers, for which there was a huge market in some parts of the world. But AeroVironment and Unique Mobility were the exceptions. Solectria continued to be the most independent small guy, but the independence was sustained primarily by DARPA's largesse.

The lesson here (if anyone cared to listen) was this: An unreliable regulatory environment created a poor seedbed for change, one with serious consequences for risk takers who relied on it for sustaining momentum. The companies could suddenly fail because the regulations died. The results? Less pressure driving new technologies into the marketplace. Big status quo companies, which could weather the vicissitudes of regulatory behavior, vicissitudes they had fanned, gained a stronger grip on technological progress. In other words, the big companies got rewarded for having successfully fought change, while those who had risked everything to make change happen got penalized. A weak regulatory environment thus bred technological authoritarianism.

Putting technology's power in the hands of a few who could control decisions about introducing the new into society and into the marketplace made them the future's pilots. The reasons they gave for either haste or delay in introducing the new were, in a capitalistic culture, motivated primarily by profit rather than by any calculation of benefits to the many. As Gregory Davis wrote in *Technology—Humanism or Nihilism*: "Never before have so few been able to bring about so much damage to so many." Lester Thurow, former dean of the Sloan School of Management at MIT, felt that the American system lacked the ability to invest in its own future. "Capitalism is myopic," he wrote in a book review of *One World, Ready or Not: The Manic Logic of Global Capitalism,* by William Greider. The system wouldn't spend on its long-term needs, from education to infrastructure building, to ensure its own survival. Capitalism, wrote Thurow, "needs government help to make those investments, but its own ideology won't allow it either to recognize the need for those investments or to request government help. That is the ideological paradox of our time."

Though imperfect and beset with human complexities, regulatory policy, if consistent, held out promise to bring new players into advanced technology arenas, which capitalism hyped as its great new frontier. A strong, clear, consistent regulatory environment could also be a fertile economic one, according to a study published by the Institute for Economic and Environmental Studies at California State University in Fullerton in 1995. The study showed that the economy of Los Angeles, with the strictest air regulations in the nation, had grown faster than the economy of the nation at large since the mid-1960s, when regulations had first gone into effect.

In 1996, despite the parade of new green cars coming down the road, no such consistent support from corporate America for regulatory policy to ensure clean air, or curb global warming, was in sight, which was ironic because regulatory pressures had been crucial for early funding and development of the technologies in clean cars; now regulators weren't sharing the glory. Maybe Howard Latin needed to add a ninth rule of bureaucratic behavior: When regulators do instigate positive change, the system sees to it that they seldom claim or receive any credit.

At any rate, at the symposium in Atlanta in late 1995, what had been in doubt the year before—that the Big Boys were in charge of clean-car momentum, their ideologies and technologies intact—was now undeniable.

"I hear they're gloating," said Paul Miller, a research fellow for the W. Alton Jones Foundation, which had funded EV projects for independents and clean air groups.

"The mandates are going down the tubes," Sheila Lynch added derisively.

"Don't get me going," Ike Bayraktar, AeroVironment's sales chief, said about the mandate. "It's eating my heart out."

As for EVs entering the American market any time soon, Bayraktar added, "General Motors is going to do it. They don't give a shit about the mandate. They don't give a shit about the government. In fact, they are like a little boy—you tell me to do it, I don't want to."

On December 22, the California Air Resources Board virtually folded on the mandate. The board's chairman, John Dunlap III, said his staff had recommended that the Big 7 be allowed to introduce ZEVs voluntarily between 1998 and 2003. No 2 percent rule in 1998, no 5 percent rule in 2001. The 10 percent rule for 2003 would stay in place, however. Defending the retreat, Dunlap declared, "We will not surrender a single pound of emissions with this plan."

Two weeks later, at the Los Angeles International Auto Show, where Roger Smith had first unveiled the Impact six years earlier, his successor, Jack Smith, announced that GM would make and sell the car, renamed the EV1, in the fall. Around the world, here was the new ripple: the biggest of the Big Boys was going to put an EV in the California marketplace. Would that mean excitement, competition, and automakers scrambling to get their green cars into production to stake out positions in an automotive revolution? Would this be the death knell of Sunrise?

19

Getting Sunrise into the Mainstream

Andrew Heafitz

James Worden and Max Schenk carrying a lightweight composite body in white for Sunrise, October 1996.

The whole face and heart of Solectria are entirely different; we feel a lot more stable and professional.

—James Worden, interview, March 1997

Maybe . . . then maybe not.

In Wilmington, James Worden was reluctantly accepting the fact that reinventing the car, leading its next big adventure, was a feat the Big Boys were going to pull off, though they—or at least one of them—sure could use Solectria's help. "The big automakers could use us," Worden said early in 1996. "But sometimes I think they're afraid of us." He now envisioned a joint venture, one to which Solectria would bring its advanced EV technologies and composite-building know-how, and the partner would bring money and assembly capability. His goal was to split ownership as close to fifty-fifty as possible and to get Sunrise more mainstream as a real car that both fleets and individuals could buy. "The bottom line," he said, "is that there has to be a shared vision to get Sunrise into production. We're not going to sell out."

Worden was tight-lipped about who such a joint venture partner might be. But he did say he had been talking with GM and with some foreign companies. Sheila Lynch at the Northeast Alternative Vehicle Consortium said her organization hoped Sunrise got made in the Northeast. "A lot of people here really want the project to succeed," she said.

One plus for Sunrise was the National Institute for Science and Technology grant. Solectria received the $6.8 million in early 1996. Having been a reluctant applicant, Solectria had Lynch to thank for her persistence at getting the company to apply. The money would push low-cost, high-volume composite manufacturing methods. The grant did have some strings attached, though. One was, as Worden put it, "a New England thing." The New England thing was that Sunrise had to be made there.

"The New England thing is a little bit like an albatross," Worden said.

A benevolent albatross, according to Lynch. Without it there was no $6.8 million. "James needs to realize the Big 7 are coming with cars," Lynch said.

They certainly were.

Evidence of the Big Boys' presence in the EV industry continued to grow. Chrysler, a company that had loudly and continually mocked EVs, was a major sponsor of the 1996 American Tour de Sol in May. Chrysler didn't race vehicles, but it hauled the ESX, a sleek hybrid that combined electric and gasoline engines, the EPIC EV van, which the company planned to sell in 1997, and the Prowler, a new entry in the ICE muscle-car sweepstakes, from site to site in an orange eighteen-wheeler.

GM also appeared on the tour, though somewhat surreptitiously. In Chesapeake City, a stop on the route, a red Impact glided silently out of nowhere, or so it seemed (a tow vehicle probably dropped it off somewhere out of sight), passed through the narrow, shop-lined streets, parked by the dock adjacent to the Chesapeake and Delaware Canal, and became one more EV that the curious could check out. Toyota didn't race or slip into one of the display spaces, but it did put a RAV4 EV on display at pretour festivities in Manhattan. The RAV4 EV was a small sport utility that Toyota would start leasing in 1997. It came with advanced batteries and had a range of over 100 miles.

In the 1996 Tour de Sol, James Worden set a new distance record, 374 miles on a single charge, in Sunrise P1, powered by Ovonic nickel-metal hydride batteries.

In early 1996, Honda and Nissan also announced that they would have EVs ready for California soon. Global computer chip giant Samsung Electronics said it was leaping into the EV industry big time, with billions of dollars and a new technical center in which EV research and development would be key elements. Samsung's move lent credence to what Sheldon Weinig had alluded to in his speech at EVS-12 two years earlier; that is, electronic component makers with vision, the piece makers that would fill the personal-mobility boxes people drove in the twenty-first century, might soon be entering the market as manufacturers rather than just suppliers.

European manufacturers weren't setting still. In the spring of 1996, German automaker Mercedes-Benz unveiled a fuel-cell-powered sedan. Mercedes said it hoped fuel cells might one day even power the Smart, the environment-friendly city car going on sale in Europe in 1998. The Smart, created in a process James Worden might learn from and admire, with Swiss watchmaker Swatch joining Mercedes-Benz, was a very funky little car indeed. Its exterior was armadillo-like; composite panels could be exchanged like chinking if the owner so desired. The Smart's superefficient ICE engine would get about eighty miles per gallon, and there was an EV version on the drawing board. Volvo and Audi were working on hybrids. Peugeot was selling its Peugeot 106 EV.

It was beginning to look like Worden's remark of August 1994—"Once there are cars being made for California, they'll be made everywhere"—was about to be validated. But how would Sunrise do in this league?

Without a doubt, the concept that had been so promising in 1994—that a new alliance of a small upstart (Solectria), a public utility (Boston Edison), the military (DARPA), and a supporting cast of small players could create the quintessential clean car a new way—had chilled. A number of things had seen to that: the crippling of the ZEV mandate, Boston Edison's getting cold feet, Worden's cautiousness in facing incongruity; the Big Boys' entrance into the fray. On the other hand, in generating momentum behind cleaner cars and their technologies, Solectria, boosted by its partners, had played a role vastly disproportionate to its size. It held out promise to continue doing so if the short-term partners stayed on board and a good long-term partner could be found.

Environmentally, of course, it mattered less which company got the first successful EV on roads in America and around the globe than when and in what numbers. Regional air pollution and global climate change were progressively more alarming. Megacities and the ICE were making sprawl, congestion, and smog more negatively synergistic. In 1996, Cairo, Santiago, Hong Kong, Athens, Beijing, and other cities were regularly putting their populations on red alert about bad air; hospitals overflowed with wheezing children and coughing seniors. "Asia Stinks" read a headline of an artcle in *Fortune.* Air quality was better in the United States, where cars at least had pollution controls, yet in and around Washington, D.C., smog had become so worrisome that TV stations were broadcasting ground-level ozone maps as part of their evening weather reports. In Los Angeles, ironically, air quality was improving. In 1994 the city had violated federal smog standards on only 100 days, a big improvement from twenty years earlier, when the first emission-control devices had appeared on new cars. Back then, in 1974, the city had been in violation on 200 days.

Global climate change continued to be hotly debated. The Clinton administration, over objections from big industry, came out in support of reductions in greenhouse gas emissions, primarily carbon dioxide. "We believe that circumstances warrant the adoption of a realistic binding target," said U.S. Undersecretary of State Timothy Wirth. Action was slow, however, and throughout 1996 anxiety about global warming fired controversy around the world. Higher annual temperatures, severe storms, and prolonged droughts, whether or not directly associated with global climate change, began to be associated with it in the media. Collectively, a universal voice of anxiety seemed to be gathering force, with European nations becoming unified in their criticism of the United States for stalling about doing anything. The central question of what would be a protracted and contentious debate was well put by Michael Oppenheimer, a scientist with the Environmental Defense Fund. Talking about the nebulous, troubling, possibly irreversible impact of global climate change, Oppenheimer said, "How can you gamble when we only have one earth to play with?"

One way was well established: deny the validity of the science that confirmed global warming; insist on more studies; get industry involved in forging distant solutions. In other words, stick with the strategy of least feasible retreat, which had proved so successful in slowing air pollution regulations in the republic for the last fifty years. If the strategy had worked so well in America, and America was the global pacesetter and standard-bearer at the end of the twentieth century, shouldn't it work for the world?

In the emerging global economy, maybe not. Nations like China and India were showing small signs of facing the intimidatingly complex personal mobility equation that development dropped in their laps. Both New Delhi and Shanghai were considering phasing out high-polluting two-stroke-engine scooters and motorcycles, providing the manufacturers of their EV equivalents a big opportunity. For countries with large populations in low-lying seacoast areas, there was more scientific evidence about climate change for both rich and poor to worry about all the time. For instance, the Antarctic ice shelf seemed to be disintegrating. In late 1996 and early 1997, an iceberg as tall as the Empire State Building and as large as Rhode Island broke free and floated toward the Pacific Ocean. Deep holes and cracks miles long were spreading through the extant ice shelf. If, as some scientists predicted, the shelf collapsed soon, it would escalate the rising of the world's oceans.

Meanwhile, looking to the United States, foreign countries saw little leadership. The ZEV mandate, though it had gotten the clean-car movement rolling, was virtually dead; the California Air Resources Board insisted that the mandate was still going to be enforced in 2003, but few people believed that. The United States also continued to drag its feet on committing to a plan for reducing greenhouse gas emissions that cause global warming. In 1997, mainstream environmental groups would begin to publicly criticize Vice President Al Gore, whom President Clinton usually deferred to when the issue involved air. "What we need and what we expect is leadership," said Deb Callahan, president of the League of Conservation Voters. Added Philip Clapp, president of the Environmental Information Center, "The failure of the White House to provide any leadership on the clean air standards and on climate change raises real questions about what real environmental progress Vice President Gore can point to in claiming the mantle of the environmental candidate in the year 2000."

The Congress didn't help much. It remained in antiregulation mode. Nevertheless, the EPA's administrator, Carolyn Browner, kept the pressure on for tougher ozone and particulate air standards across the nation. Having lost the ozone transport battle in the Northeast, perhaps the agency was hoping that by enlarging the scope of the problem it might gain enough support to push stronger regulations into law. The higher standards had industry, which insisted they would cost billions and have negligible health impacts, up in arms. Testifying before a congressional subcommittee in defense of the standards, Browner said that her agency had considered more than 240 peer-reviewed scientific studies—the most ever considered by the EPA for a health standard review. "The current levels leave too many people at risk," Browner said. "That's what the science shows us. Too many people are suffering."

Sunrise, despite its small scale, was a beacon of enlightened possibility in this picture. As prototype P3 came together, Sunrise had promise; it had glow. It was rolling proof of the technological potential of clean cars to actually make a big dent in the personal mobility equation soon. If someone would make it and, of course, sell it.

Putting Some Thunder in the Sunrise Program

In the spring of 1997, Solectria was in new and larger quarters. The company now occupied a stand-alone building in the same industrial park in Wilmington. The engineers had painted the new place, wired the phones and computers, and set up the machine shop, electrical testing room, conversion space, and offices. Sunrise had undergone an upgrade from a one-room operation to three workrooms, for carpentry, mechanicals, and composites, and its own small office space.

Sunrise also had a new program director. Jeff Fisher had come on board the previous October, replacing Mark Dockser. Fisher was a little older than the typical Solectria wunderkind and had a broader background. He'd worked in finance, electronics, and planning. He'd spent stints at IBM, the World Bank, and the Harvard MBA program. Just before coming to Solectria, Fisher had worked at H Power Corporation, where he'd raised capital, written business plans, and managed projects, including one for a $16 million fuel cell/battery hybrid bus. Getting up to speed on Solectria and Sunrise, Fisher found himself a little overwhelmed. He put in fourteen-hour days, wrote an eighty-page business plan, and attempted, in his words, "to put some thunder in the Sunrise program." By early 1997, Fisher was able to focus on his main reasons for being at Solectria: finishing the development of Sunrise and reeling in the right strategic joint partner to make production real.

Other talented newcomers had joined the Sunrise team as well. They included platform engineer Michael Jones, a Brit with the accent to prove it. Like Fisher, Jones had broad experience—in his case, automotive experience with small carmakers like Jensen and Sigma and Lotus. At Lotus, an elite British marque, he had been project director for the Elan, a composite-body sports car. Jones had gray hair. Vasilios Brachos, thirty, was lead structural design engineer. He'd been involved with Sunrise while at the University of Lowell, which continued its relationship with Solectria as the crash-test partner, but now worked full-time in Wilmington. Brachos's challenging job was to simplify and optimize the use of composites in Sunrise; he and safety systems engineer Ken Sghia-Hughes cooperated to make the car strong, light, and safe. Stylist Clark Taylor's job was to adapt James Kuo's aesthetics to accommodate evolving engineering demands. In addition, the distinct component systems, such as brakes and suspension, heating and air conditioning, each had a lead engineer. In other words, Sunrise was beginning to resemble a real, albeit small, design group in the real automotive world. Its staff numbered about fifteen, mostly engineers. This was far from what even a low-volume carmaker such as Lotus had fielded, noted Jones. "At the end of the development program at Lotus," he said, "we had about 400 engineers."

The near-term goal of the Sunrise team was to finish prototype number three. When P3 was done, said Wayne Kirk, who continued as the managing engineer for the body, "if it really works, we'll do eight or ten more for the proof of concept; if it doesn't work, we'll do one off again."

Getting P3 right would wrap up the design for manufacturing study. The car would be improved all around: suspension, interior, electronics, finish. With one good manufacturable prototype as a template, the so-called bodies in white, the phrase used to describe the lightweight composite body shells, could be made and fitted with systems. Then each car would be tested and fine-tuned. Several would be crashed once other tests were completed. As these proof-of-concept cars were being built between mid-1997 and late 1998, parallel research, funded by the $6.8 million NIST grant, was ongoing to get the costs down on composite body production processes. Even in early 1997, though, Fisher said, "With composites, Solectria is getting where we know stuff that other people don't know."

DARPA was the main funder of the proof-of-concept cars. Adept at keeping the cash flow coming from a tightfisted Congress, DARPA justified its ongoing support of Sunrise, both for the all-around upgrade of P3 and for the final proof of the concept, as a way to validate a five-year investment in the far-ahead-of-the-PNGV-pack Sunrise and in youthful upstart Solectria. According to Sheila Lynch, the funding for P3 and the concept vehicles would probably be the last money Solectria could expect to see from the military.

That was okay with James Worden. He wanted Solectria weaned off government funding. In mid-March of 1997, a little older, with lines around his eyes eliminating the chances he'd be carded, a little more subdued, he said, "Commercial voices are the ones we need to hear now. The reality of those voices is what we need to hear. We'd like to be forced to put Sunrise into production. We'd like the forces to come from the right partner. Sure, there might be some hard things we don't like. But they won't be the forces of the past—things out of the blue—this dreamer and that dreamer, this political force and that one. Where nothing makes any sense and we're all pulling in different directions."

Worden mentioned EV1, which had been launched in southern California and Arizona in December of 1996. Despite an expensive campaign for the launch and extensive personal service through Saturn dealers, who handled the cars, leases had been fewer than expected. Some critics said the lease price was too high, the marketing campaign unfocused, the lead-acid batteries inadequate for the range even early adopters wanted. Worden, though, was complimentary. "It's new and it takes time," he said. "We've been doing this for five years, with hardly any marketing except word of mouth. GM has leased 160 cars in four months. It takes a while to gain momentum." As for expanding the market, he went on, "GM is looking anywhere that's warm and flat. And we started with cold and hilly . . . and those are the places we're going to be looking to enter the market. Cold and hilly allows us to think more globally."

Global marketing of Sunrise, in either cold and hilly or warm and flat regions, was still a few years away. "The market is global, though," Jeff Fisher emphasized. "We have certain vehicles and components now under evaluation by worldwide automakers."

Still, Michael Jones, was quick to point out, "We're still doing one-offs. No way can we get into production with this small team. We're really dependent on partnerships, where we'll have access to more resources, more facilities, and we can go into production."

Walking through Sunrise's three-room work area, which he designated as "dirty, clean, and cleanest," for carpentry, mechanical, and composite work, respectively, Jones pointed out instrument panel pieces from Textron, Inc., being fit into P3. A couple of technicians examined a prototype door. "There's a lot of engineering in a door," Jones said. A body in white from TPI Composites sat on the floor, white as cream, smooth as marble. It looked as much like an aesthetically pleasing piece of sculpture as it did the body of a futuristic car. "The major design effort has been on the body here," Jones said. "The next major effort has been in transferring the electrical goodies that Solectria has developed, like the motor, like the controller, all that stuff that's part of the vehicle, and our battery technology, along with other components that have to be modified, onto the body. All the money that's coming in from outside is for the body. The mechanical parts at the moment are only a way to demonstrate the body."

Asked what he liked best about working at Solectria, Jones said quickly, "We don't have to issue memos." More seriously, he went on, there was a "can-do" attitude at this company unlike any he had seen elsewhere. "These guys will do anything." As an example, Jones told the story, obviously a piece of the growing Solectria folklore, of electrical engineer John Rogers reinventing the Sunrise accelerator. Originally, it shook and produced sudden lurches that drivers weren't prepared for. Mulling over the problem in the shower ("I always get my best ideas there," he explained), Rogers later went into work and redesigned the accelerator pedal in an hour. It no longer shook.

"We don't know that it can't be done," said Rogers happily, recalling that day. "I love it here. This is one of the few places in America you can be really creative and people will support it."

Epilogue

In late October 1997, Worden drove a Sunrise prototype, dubbed P2001, from Boston to New York, using less energy than it took to cook the typical American Thanksgiving dinner. The publicity stunt helped Worden receive a merit award from CALSTART for advanced transportation leadership, but did not get Sunrise any closer to a deal with a big partner eager to manufacture it. "Even if you've got a better car," said Sunrise program director, Jeff Fisher, "the automakers want to use their own research and development."

In early December, Solectria took Sunrise to EVS-14, held at Disney World in Orlando. The car had lots of company. In addition to buses, trucks, and vans on display, automakers had more than a dozen EVs available for ride-and-drives, with most of them also on sale. Toyota led the pack with its RAV4 EV, its Prius HEV, which already had orders for over 1000 vehicles in Japan, and its e-com, a lightweight, two-passenger EV with a sixty-mile range and a price tag of between $12,000 and $14,000. Among the smaller players, Unique Mobility got plenty of attention with its Ethos 3 EV, a joint venture with Pininfarina, the Italian bodymaker. Unique Mobility also introduced another new partner, former Chrysler CEO and super salesman, Lee Iacocca. Like former GM CEO, Robert Stempel, Iacocca had glimpsed the future. And it ran on electricity. As a result, he was backing a new venture, EV Global Motors, to make electric scooters for Asia. Unique Mobility was supplying drivetrain components for the niche-market scooters.

Solectria also made a new vehicle introduction at EVS-14. The company rolled out its CitiVan, an ugly but practical delivery vehicle aimed for what Worden called "a niche within a niche market." That is, an electric van for the already limited van market, a market that the major automakers ignored because of its small size.

Solectria had to do something. Sunrise wasn't taking off as hoped. As predicted, sales of the company's workhorse, the Force, were declining now that automakers had quality EVs available. The company's E-10 pickup had been discontinued. So launching a niche vehicle as a short-term survival tactic was smart. Solectria was enjoying an increase in sales of components to bus and van manufacturers, so doing a van of its own was a natural.

DARPA and Boston Edison, which had patched up its relationship with Worden and the Wilmington gang, were bankrolling the CitiVan. NAVC's Sheila Lynch was coordinating the effort. At EVS-14, Solectria got eleven orders for the vehicle. Production would begin in mid-1998 in a facility provided by Boston Edison.

Throughout 1997 global climate change received tremendous attention in the press and was the focus of a conference held in December in Kyoto. There, representatives from more than 150 nations met to negotiate an international accord on how to lower greenhouse gas emissions. The Kyoto conference put some heat on the United States, the 10-ton gorilla at the event, since it produced almost a quarter of the world's carbon dioxide emissions. In 1992, at the Rio de Janiero "Earth Summit," the United States had refused to even consider voluntary reductions in greenhouse gas emissions. In Kyoto, despite strong opposition from a coalition of American industries, the United States did take a stand and agree to reduce carbon dioxide emissions to 7 percent below 1990 levels by the years 2008 to 2012—if Congress ratified the treaty.

In early 1998 President Clinton, who publicly supported the Kyoto accord, introduced a tax incentives and research package for clean cars. Critics didn't like it; they said that a decision on the Kyoto accord was needed before tax credits were offered in support of its goals. Senator James Inhofe (R.-OK) accused Clinton of trying "to mold the behavior of U.S. businesses to conform with the global warming ideology."

Ideology? Fewer and fewer scientists thought so. Evidence continued to accumulate, from ice core drillings in Antarctica, from sophisticated computer models that integrated complex variables, from temperature data, that atmospheric change was zealously afoot. Speaking for this group, journalist Ross Gelbspan, in his book *The Heat Is On,* asked, "Do we need more evidence that climate change is affecting the planet? The human population has become so huge and our technology so powerful that the weight of our industrial emissions—amplified by the cutting down of CO_2-absorbing trees and forests—is having impacts on the earth in ways we never intended. We are twisting our planet out of shape. We didn't mean to, but we are."

Auto and oil interests disagreed. The science behind global warming remained shaky, they said; the answers to key questions were not in. Until they were in, drastic change of the status quo was economically dangerous.

This was the same reasoning the industries had used for decades to stall regulatory change to reduce ground-level ozone. If the oil and auto industries had their way, no consensus on global warming would be achieved until New York was hubcap deep in the Atlantic Ocean. Still, one oil CEO did break ranks a little. British Petroleum head John Browne said publicly that there did seem to be a link between greenhouse gases and global warming. "It would be unwise and potentially dangerous to ignore the mounting concern," he admitted.

Meanwhile, air pollution plagued more and more cities worldwide. In late 1997 Paris and Moscow had major smog alerts. Jerusalem's air would rival Mexico City's by 2010, claimed an article in the *Jerusalem Post.* Florence, Italy, banned older vehicles; Shanghai limited the number of gasoline scooters. In the United States the EPA began publishing an ozone information map, updated daily, on the World Wide Web. The agency also printed a report on the environmental benefits of the Clean Air Act between 1970 and 1990. According to the report, benefits exceeded costs to achieve them by a ratio of forty to one. Monetary benefits totaled $22 trillion, while costs were $523 billion.

Global warming and air pollution notwithstanding, worldwide sales of ICEs were increasing, especially in developing countries. Nightmarish traffic jams in cities around the world were practically routine. From Kuala Lumpur to São Paulo, from Bangkok to Istanbul, from Djakarta to New Delhi, traffic never slept, nor did emissions cease. Frequently, public transit systems withered and development patterns reliant on the car reshaped city centers and suburbs and swallowed up surrounding countryside in a model created in America.

Despite the gridlock and immense emissions, automakers were selling more cars and trucks and buses. Between 1997 and 2000 three quarters of the new auto factories coming on line were located in developing countries, where there seemed to be an insatiable hunger for cars. In developing countries oil companies were erecting gas stations and delivery systems as fast as they could to refuel the traffic in spite of the fact that even conservative estimates gave increasing petroleum production a very short life span. If present usage patterns continued, oil production around the world would peak, and start to taper off, with prices escalating accordingly, by 2020. New ways to extract oil would be developed but they would be expensive.

Nineteen ninety-seven proved to be the warmest year on record, eclipsing 1990 as the warmest to date. Nineteen ninety-seven also proved to be the year the ZEV mandate got completely killed. The mandate had arguably moved clean-car developments ahead faster than anything else. New technologies, new clean cars, and new attitudes had resulted from the mandate. In late 1997, though, it survived in only one state, New York. Like rogue white blood cells in an immune system gone haywire, industry lawyers clustered on the last regulatory clot to rid it from the body of the transportation system. Soon they were successful.

Automotive corporate behavior remained adolescent: ego-driven, focused on short-term gain, wildly optimistic despite clamorous warnings. Any teenager, or CEO, can deny the probable realities of the future and live hedonistically; it takes wisdom, and maturity, to do otherwise. But because regulations are rules, they chafe the corporate adolescent spirit. Getting together to kill the rules is an act the Big Boys can do together. They can be an antiregulation team of immense savvy, strength, and persistence. After each victory, members can return to their respective dugouts in Detroit and get on with their oligopolist game of business, as they like it played.

With the ZEV mandate dead and the Big Boys in control of EV developments, where did that leave the upstart start-up Solectria in the spring of 1998?

Basically, it left it with its conversion vehicles dropping out of production, with a delivery van going into production, with component sales strong, and with Sunrise still seeking the right big partner to make it real. "We get orders for Sunrise," Jeff Fisher said, "we're ready for a moon shot."

Notes

1. The Freewheeling James Worden Wins Another Race

3 **"James is a genius":** Sheila Lynch interview; James Hogarth interview.

3 **"James is the Bill Gates of EVs":** Bill Van Amberg, CALSTART, in conversation.

3 **"James is a nerd":** C. Michael Lewis interview.

4 **Lin Higley quotes:** Lin Higley interview.

4–5 **NESEA American Tour de Sol facts:** Nancy Hazard interview.

6 **"You got to be a blood-and-guts driver":** Art Liskowsky in conversation.

6 **Rules of Tour de Sol:** Nancy Hazard, Rob Wills interviews.

2. Then and Now: Rebirth of a Dormant Industry

11–12 **Charles Jasper Glidden and the Glidden tours:** Thomas H. Russell, ed., *The American Cyclopedia of the Automobile* (Thompson and Thomas, 1909), pp. 151–153.

12 **Thomas Davenport:** Michael Brian Schiffer, *Taking Charge: The Electric Automobile in America* (Smithsonian Institution Press, 1994), p. 7.

13 **Niche EVs occupied:** Schiffer, *Taking Charge;* Schiffer writes concisely about the rise and fall of EVs during the first quarter of the twentieth century.

13 **Electric car disappeared:** Schiffer, *Taking Charge*; see chapter 11, "The Denouement," about the decline of the industry in the early twentieth century. See also Daniel Sperling, *Future Drive* (Island Press, 1995), p. 36; Sperling writes: "By 1910 the heyday of the electric runabout was over. The Ford Model T was now selling for less than half the price of any advertised electric car. By 1915, less than 2 percent of the 2.5 million motor vehicles in operation in the United States were powered by electricity. The electric industry dwindled away, with the last factory in the United States closing in 1935."

14 **Stempel contrasts ICEs to EVs:** Speech by Robert Stempel at the Twelfth International Electric Vehicle Symposium, December 1994.

15 **New light:** Daniel Yergin, *The Prize* (Simon and Schuster, 1991), p. 14: "In the first decades, the oil business provided an industrialized world with a product called by the made-up name of 'kerosene' and known as the 'new light,' which pushed back the night and extended the working day."

15 **"An unmitigated nuisance":** John Bentley, *Great American Automobiles* (Prentice-Hall, 1957), p. 153.

15 **"For no earthly reason other than":** Bentley, *Great American Automobiles,* p. 154.

16 **"In the absence of signposts":** T. R. Nicholson, ed., *The Motor Book: An Anthology 1895–1914* (London, 1962), pp. 267–268.

16 **"The blood-red trail of the Gliddenites":** Nicholson, *The Motor Book*, p. 269.

16 **Packard driver died:** Nicholson, *The Motor Book*, p. 268.

17 **End of Glidden tours:** American Automobile Association. The AAA took over the tours in 1908, running them until their termination in 1914.

3. James Testifies on the Promise of Clean Cars

19 **James Worden quote:** James Worden interview.

19–20 **Testimony before congressional subcommittee on energy:** From statement by James Worden made to subcommittee on June 30, 1994.

20 **James Worden quotes:** Statement to Subcommittee on Energy.

20 **Electric vehicles on cover of *Business Week:*** David Woodruff, "Electric Cars: Are They the Future?" *Business Week*, May 30, 1994, pp. 104–114.

21 **David Cole quote:** David Cole interview.

4. An Idea Whose Time Has Come?

23 **Big 3 profits:** *Automotive News,* February 6, 1995, p. 1.

23 **Four largest small companies:** Though statistics were hard to come by for the small guys, David Woodruff ("The Not-So-Big Wheels Leading the Charge," *Business Week*, May 30, 1994, p. 114) listed some of the little players: Solectria's 1993 revenues were $2 million, its earnings not available; Renaissance Cars' 1993 losses were $.54 million; Unique Mobility's losses were $2.5 million; and U.S. Electricar's losses were $2.6 million. The two other EV start-ups mentioned were AC Propulsion in California, with six employees and $.7 million in revenue for 1993 (no earnings figures available), and Rosen Motors (funded by Ben Rosen, the chairman of Compaq Computer Corporation), with no numbers of any kind available.

24 **Big 3 just emerging from a tough decade:** Lindsey Chappell, *Automotive News* reporter, in conversation.

5. James's Second Car

27 **Solectria I and II:** James Worden, John Worden, C. Michael Lewis interviews.

27 **Worden first meets Ed Trembly:** Ed Trembly interview.

28 **Solectria III:** James Worden, Anita Rajan interviews.

28 **Swiss Tour de Sol:** C. Michael Lewis, Robert Wills interviews.

28 **Bike technologies:** James Flink, *The Automobile Age* (MIT Press, 1988), p. 5.

29 **"We were scared to death":** John Worden interview.

6. Racing Solar-Electric Cars at MIT

31 **Catherine Anderson:** Anita Rajan, Catherine Anderson interviews.

31 **Formation of MIT solar racing team:** Anita Rajan, Catherine Anderson interviews.

31 **Not getting official MIT club status:** James Worden, Catherine Anderson interviews.

32 **Anderson quotes:** Catherine Anderson interview.

32 **Guerrilla engineering:** Catherine Anderson, Anita Rajan, James Worden interviews.

33 **Appropriate acts of contrition:** Catherine Anderson interview.

7. Air

35 **"We were more technology junkies":** Catherine Anderson interview.

36 **"This most excellent Canopy, the air":** William Shakespeare, *Hamlet*, act 2, scene 2; the monologue goes: "This most excellent canopy, the air, look you, this brave o'erhanging firmament, this

majestical roof fretted with golden fire, why, it appears no other thing to me but a foul and pestilent congregation of vapors."

36 **Smog and the story of auto air pollution:** James E. Krier and Edmund Ursin, *Pollution and Policy: A Case Essay on California and Federal Experience with Motor Vehicle Air Pollution, 1940-1975* (University of California Press, 1977), is a thorough, invaluable guide. Randall B. Ripley, "Congress and Clean Air: The Issue of Enforcement, 1963," in *Congress and Urban Problems* (Brookings Institution, 1969), is a good look at early attempts to pass laws to deal with air pollution. See also Arthur C. Stern, "History of Air Pollution Legislation in the United States," *Journal of the Air Pollution Control Association,* January 1982, pp. 44–61. In addition, various Environmental Protection Agency bulletins detail the history of individual pollutants, including efforts to scientifically identify and abate them.

37 **Air science since 1950s:** Beginning in the late 1940s, California led the way in discovering how pollutants reacted with air, people, and things. In 1955 the federal government funded a five-year research and training plan that states could enroll in voluntarily. The bill was extended, but during the 1950s and early 1960s the consensus was that air pollution was not a federal problem—it was a state problem. The thinking started to shift in 1963 when the Kennedy administration pushed through the original Clean Air Act. In 1967 the federal Air Quality Act set standards for all states except California, which had already set more rigorous ones.

36 **Smog upstaged by new contenders:** *The EPA's Plain English Guide to the Clean Air Act* (EPA, 1993).

36 **Smog movement and NOx scavenging:** Richard Poirot interview.

37 **Smog misnamed:** Krier and Ursin, *Pollution and Policy*, p. 1.

37 **VOC and NOx emissions:** James J. MacKenzie, *The Keys to the Car: Electric and Hydrogen Vehicles for the 21st Century* (World Resources Institute, 1994), p. 3.

38 **Least feasible retreat:** J. Esposito, *Vanishing Air* (Grossman Publishing, 1970), p. 41; called "minimal feasible retreat" in Esposito's book. Krier and Ursin describe the key features of the strategy (*Pollution and Policy*, p. 89).

38 **Haagen-Smit and his early research:** Gladys Meade interview; Krier and Ursin, *Pollution and Policy*, pp. 77–90.

38 **Ozone burn:** Catherine Anderson, comment on draft manuscript.

38 **Trolley lines:** Jim Klein and Martha Olson, "Taken for a Ride," television documentary, 1996.

39 **Scientific studies:** Krier and Ursin, *Pollution and Policy*, p. 93.

39 **Los Angeles's atmospheric peculiarities:** Krier and Ursin, *Pollution and Policy*, pp. 41–52; Cabrillo "noticed that although mountain peaks were visible in the distance, their bases were obscured"; the 1903 editorial said the air was so thick that traveling "was like meeting a railroad train in a tunnel"; the city's inversion layer is "a transparent sheet of air extending over the entire Los Angeles area, at a level usually varying between 1,000 and 3,000 feet. Sometimes such inversion layers are so sharp and definite a stratification that a balloon, ascending slowly, will rebound momentarily from their under surfaces."

41 **Haagen-Smit's battles:** Gladys Meade interview; Krier and Ursin, *Pollution and Policy*, chapter 6, "Discovering and Documenting the Role of the Automobile."

42 **Complex studies of smog:** Those mentioned, along with dozens more, conducted under auspices of the Health Effects Institute. "Estimation of Risk. . .", 1985, by Marie A. Amoruso, and "The Role of Ozone. . .", 1992, by David G. Thomassen.

42–43 **Proof of smog's harm difficult to assess:** Between 1982 and 1990 the Health Effects Institute in Cambridge, Massachusetts, a research organization focused exclusively on the effects of automobile exhausts, spent 40 to 50 percent of its research funds on ozone and nitrogen dioxide, both oxidants associated with diseases, with little conclusive proof of harm. On other hand, in 1994 the American Lung Association claimed that the EPA's standard for ozone (120 parts per billion) was unsafe and was suing to have the standard lowered.

43 **"The answer is not in":** Jane Warren interview.

43 **Ozone decreases lung function:** H. Gong Jr., "Health Effects of Air Pollution: A Review of Clinical Studies," *Clinics in Chest Medicine (*1992): 201–214. For how hospital admissions for respiratory and cardiac disease increase in relation to ozone levels, see J. Schwartz, "PM 10, Ozone, and Hospital Admissions for the Elderly in Minneapolis–St. Paul, Minnesota," *Archives of Environmental Health* 49 (1994): 366–374.

43 **Sherwin study:** Shari Roan ("Air Sickness," *Los Angeles Times*, April 3, 1990, pp. E1, 11); summarizes Sherwin and his assistant Richert's study, which Sherwin presented at the Conference on Tropospheric Ozone and the Environment, Los Angeles, March 1990.

43 **Alveoli:** John W. Hole Jr., *Human Anatomy and Physiology* (Wm. C. Brown, 1990).

44 **Number of autos, emissions increasing worldwide:** Daniel Sperling, *Future Drive* (Island Press, 1995); Deborah Gordon, *Steering* a *New Course: Transportation, Energy, and the Environment* (Union of Concerned Scientists, 1991); see also Michael Renner, *Rethinking the Role of the Automobile,* Worldwatch Paper 84 (Worldwatch Institute, 1988), for a succinct look at the increase in vehicles around the globe since 1950.

46 **In U.S. overall emissions down:** *National Air Quality and Emissions Trends Report, 1994* (EPA, 1995), pp. 1-1, 1-5; "Emissions for all (six EPA) criteria pollutants except nitrogen oxides decreased between 1970 and 1994," the report says. "While progress has been made, it is important not to lose sight of the magnitude of the air pollution problem that still remains. . . . About 62 million people in the United States reside in counties that did

not meet a minimum of one air quality standard based on 1994 monitoring data."

47 **Lifestyle changes remained taboo:** Richard Brooks interview.

8. Australia, 1987

49 **"It's in my blood":** James Worden interview.

49 **"It was quite a saga":** Catherine Anderson interview.

50 **World solar challenge:** William Jordan Jr., "Powered by Sunlight, Alien Autos Test Themselves for 1,950 Miles," *Smithsonian*, September 1988, pp. 49–58. Also James Worden, Alec Brook interviews.

50 **Solectria IV fire:** Catherine Anderson interview.

51 **"It wasn't any wonder":** Howard Wilson interview.

9. 1989

53 **Anita Rajan:** Anita Rajan, Catherine Anderson interviews.

54 **Paul B. MacCready Jr. and Aerovironment:** AeroVironment brochure.

55 **Setting speed record:** James Worden interview.

55 **Anita Rajan quotes:** Anita Rajan interview.

56 **Anita Rajan quotes:** Anita Rajan interview.

10. Forces Gathering Behind Cleaner Cars and Air

59 **Economic Growth/Clean Air Paradox:** Jananne Sharpless interview. K. T. Berger, *Where the Road and the Sky Collide* (Henry

Holt, 1993), part 1, provides a good overview of the California situation. CARB Technical Support Documents for the 1990 ZEV mandate statistically document the conundrum of clean growth/dirty air when ICEs are the means of getting to and from work and the resultant growth is sprawling.

60 **Jananne Sharpless quotes:** Jananne Sharpless interview.

61 **Methanol study panel:** Jananne Sharpless interview.

61 **Methanol as fuel:** Gordon, *Steering a New Course,* pp. 85–88; Alan Lloyd interview.

61 **Reformulated gas:** Jananne Sharpless interview. For a discussion of reformulated gas, see Gordon, *Steering a New Course,* pp. 93-95.

61 **"The situation had the oil companies":** Jananne Sharpless interview.

62 **"I called it 'the jolt'":** Paul MacCready interview.

62 **Brief history of Clean Air Act:** Richard Brooks interview. Also, "The Clean Air Act Amendments of 1990: A Symposium," *Environmental Law* 21 (1991).

63 **"One of the longest—and hardest fought—legislative battles":** Henry A. Waxman, "An Overview of the Clean Air Act Amendments of 1990," *Environmental Law* 21 (1991): 1723–1816.

64 **Smog levels in Boston area:** Dan Greenbaum interview.

64 **NESCAUM analysis of federal plan:** Michael Bradley interview.

65 **Market research conducted by GM:** GM publication on "Impact Specifications."

65 **"When I first heard about the Impact":** Norm Salmon interview.

65 **Bill Sessa and Roger Smith exchange:** David Sedgwick and Bryan Gruley, "'You Guys Aren't Going to Make Us Build That Car, Are You?'" *Detroit Free Press,* May 7, 1994, pp. 1B, 4B.

11. 1989–1992

68 **Lightspeed:** James Worden interview.

68 **"Those days James was convinced":** Ed Trembly interview.

68 **"I sweep the floors":** Ed Trembly interview.

69 **Environmental investor Jay Harris:** Anita Rajan, Jay Harris interviews.

69 **"It was just exactly like that":** Anita Rajan interview.

69 **"Why don't you lower your sights":** Ed Trembly interview.

69 **"James wanted to be convinced":** Ken Sghia-Hughes interview.

70 **Trial conversion:** Ed Trembly, Ken Sghia-Hughes interviews.

70 **Other small companies:** In the United States the main small companies were AC Propulsion, U.S. Electricar, and Rosen Motors in California; Renaissance Cars in Florida; and Unique Mobility in Colorado.

70 **"That car howled":** Ed Trembly interview.

71 **It was the old pattern:** Anita Rajan, Ed Trembly, Scott Hankinson interviews.

71 **Worden's philosophy toward the press:** Anita Rajan interview.

71 **"I was ready to scream":** Anita Rajan interview.

71 **"I'm not looking for a real job yet":** Scott Hankinson interview.

71 **"We had a month to recoup":** Anita Rajan interview.

72 **U.S. Electricar:** John Dabels interview; U.S. Electricar annual report for 1993; David Woodruff, "Electric Cars," *Business Week*, May 30, 1994, pp. 111, 114.

72 **Renaissance cars:** Frank Markus, "The Electric Tropica," *Car and Driver*, March 1994, pp. 95–97; Markus claimed that Beaumont's Sebring Vanguard vehicle "didn't win any beauty pageants—it looked like a cross between a doorstop and a milk carton."

72 **Backlash gathering against EV mandates:** Sedgwick and Gruley, "You Guys Aren't," p. 4B.

72 **"It was an amazing year":** Anita Rajan interview.

72–74 **Zinc-bromine battery incident:** Ed Trembly, James Worden, Catherine Anderson, Anita Rajan, Arvind Rajan interviews.

12. A Regulatory Minefield

77 **Pentagon's interest in electric vehicles:** Matthew Wald, "Pentagon Turning Plowshares into Swords," *New York Times,* March 16, 1994, pp. A1, D5.

78 **Conservation efforts since the 1970s:** Amory B. Lovins and Hunter L. Lovins, "Reinventing the Wheels," *Atlantic Monthly*, January 1995, pp. 75-93.

79 **"Was the single wild card":** Rusty Russell interview.

80 **The Ozone Transport Commission:** Daniel Greenbaum, Rusty Russell, Sheila Lynch interviews. The Ozone Transport Commission itself also published briefs, texts of recommendations, and press releases.

80 **"Is a howling blizzard of bizarre matter":** David Bodanis, "Pulling What Out of Thin Air?" *Smithsonian*, April 1995, p. 76.

80 **NESCAUM studies:** Michael Bradley interview. See also "Adopting the California Low Emission Vehicle Program in the Northeast States: An Evaluation," a study by NESCAUM, September 1991; and "Strategies for Using Renewable Resources to Reduce Greenhouse Gas Emissions from Electric Vehicles in the Northeast," another study by NESCAUM in collaboration with the Consortium for Regional Sustainability, January 1994.

81 **"What's killed us in the past":** Michael Bradley quote.

81–82 **Detroit opposes northeast mandate:** Sedgwick and Gruley, "You Guys Aren't," pp. 1B, 4B.

83 **India marketplace for cars:** John McElroy, editor of *Automotive Industries*, comment to audience of Society of Automotive Analysts, Detroit meeting, January 3, 1994.

83–84 **Final vote on OTC:** Dan Greenbaum interview.

84 **Supercar project:** Matthew Wald, "Government Dream Car," *New York Times,* September 30, 1993, pp. D1, D7; White House press releases.

85 **Thomas Jorling quote:** Dan Greenbaum interview.

85 **Tom Hayden quotes:** Tom Hayden, "Conspirators Targeting Electric Cars," *EV Insider,* December 1993, p. 3.

86 **Little guys would get a piece:** Ron Chapman interview.

86 **Roundtable worries:** Sheila Lynch interview.

87 **Forty-nine-state car:** Since the first emission laws in the 1960s, California typically initiated and passed legislation first, with federal regulators often adopting variations of the legislation for the country at large. Through the Clean Air Act Amendments of 1990 and the OTC proposals, California continued to be treated separately; thus there were usually two sets of laws, one for California, the other for the remaining forty-nine states. The "forty-nine-state car" referred to clean car standards slightly different, and usually less rigorous, than those in effect in California.

87 **Sentiments in the Pro-EV camp:** Dick Mark memo, July 18, 1994, to ZEV activists and advocates campaigning for the OTC petition; Sheila Lynch interview.

88 **"Thought collectives":** William Irwin Thompson, *Imaginary Landscape* (St. Martin's Press, 1989), p. 110.

89 **"The auto industry is after delays:** Jamie Buchanan interview.

90 **"The joke is that":** Jason Grumet interview.

91 **U.S. Justice Department antitrust suit:** Paul J. Miller, "Technology-Following, Technology-Forcing, and Collusion in the Auto Industry" (W. Alton Jones Foundation, 1995), pp. 4, 5.

92 **Laws of administrative behavior:** Howard Latin, "Regulatory Failure, Administrative Incentives, and the New Clean Air Act," *Environmental Law* 21 (1991): 1649-1652.

13. "No Zero-Emission Mandate Is a Stake in the Heart"

95 **"No ZEV mandate is a stake in the heart":** Arvind Rajan interview.

95 **James Worden quotes:** James Worden interview.

95 **Anita Rajan quotes:** Anita Rajan interview.

96 **"You want a real societal mandate":** Paul MacCready interview.

96 **Amory Lovin's Hypercar:** Lovins and Lovins, "Reinventing the Wheels," pp. 75–86.

97 **"A Hypercar is far beyond what we're doing":** Anita Rajan interview.

97 **"It sounds simple. Try building one.":** James Worden interview.

14. The Sunrise Project and Its Partners

100 **Business growth:** Arvind Rajan, Mark Dockser interviews.

102 **James Kuo quotes:** James Kuo interview.

103 **Design of a new car is story of compromises:** Wayne Kirk interview.

103 **"It was a huge effort":** Wayne Kirk interview.

105 **"It's his dream project":** Norm Salmon interview.

105 **"A base of no fear":** Rob Wills interview.

105 **"The usual Solectria madness":** Ed Trembly interview.

106 **Big 3 agendas different:** James Worden interview.

107 **Phase one, phase two:** Sheila Lynch interview; NAVC materials.

108 **"Leading the military's charge":** Wald, "Pentagon Turning Plowshares into Swords," p. A1.

108 **"Spin on," "spin off":** Wald, "Pentagon Turning Plowshares into Swords," p. D5.

108–9 **DARPA in the past:** Dave Maass interview.

109 **Sheila Lynch quotes:** Sheila Lynch interview.

109 **Mutual admiration society:** Richard Watts interview.

110 **"DARPA is really the driving force here":** Arvind Rajan interview.

110 **James Hogarth career:** James Hogarth interview.

111 **James Hogarth quotes:** James Hogarth interview.

111 **A man "with grandoise schemes":** Sheila Lynch interview.

111 **"Those two-seat, rinky-dinky cars":** James Hogarth interview.

112–13 **Major Richard Cope quotes:** From Cope's pep talks at the Smuggler's Notch ARPA meeting, October 1994.

113 **"People don't want those":** Don Crockett in conversation.

114 **"I'd like to call it the Fly":** Jerry McAlwee in conversation.

115 **"A lot of people are on the steep part":** Jerry McAlwee in conversation.

115 **"I'm very much a systems person":** James Worden interview.

116 **Ford Ecostar fires:** Eventually, Ford canceled plans to produce its sulfur-battery–powered Ecostars.

116 **TNE II accident:** Author on scene.

116 **Vermont farmer:** Richard Watts interview.

117 **"The whole thing is organic":** Dave Blair interview.

119 **"Boston Edison not an investment angel":** James Hogarth interview.

120 **Solectria pickup uninvited:** Arvind Rajan interview.

15. Anaheim, 1994

126 **PrEView Impact:** GM PrEView Drive Program Fact Sheet.

126 **"If we did what Solectria did":** Sean McNamara in conversation.

127 **Reporters' comments:** Press room comments at EVS-12.

127–28 **GM wish list:** Ken Baker speech at EVS-12.

128 **Senator Barbara Boxer quotes:** Barbara Boxer speech at EVS-12.

128–29 **Sheldon Weinig recounted lessons:** Sheldon Weinig speech at EVS-12.

129 **Paul MacCready quotes:** Paul MacCready interview.

130 **"It's classic automotive technology push":** Sheila Lynch interview.

131 **"As a result of their experience":** Miller, "Technology-Following, Technology-Forcing," p. 5.

131–32 **James Worden quotes:** James Worden interview.

16. Sunrise Almost Stalls, Then Rolls, Finally Races

135 **Meeting a bust:** Sheila Lynch interview.

136 **"James Worden Is afraid to leapfrog":** James Hogarth interview.

136 **"Jim Hogarth is way off key":** James Worden interview.

136 **NIST grant:** Sheila Lynch, Wayne Kirk interviews.

137 **"James Worden should be courting us":** Dave Maass interview.

139 **"Without DARPA money":** James Worden interview.

139 **"That's a hard reality":** John Gully interview.

140 **"The archtype of the depersonalized":** Flink, *The Automobile Age*, p. 232.

142 **"They thought it was from Mars":** Jim Ellis interview.

142 **"Our hope is that China, India":** James Hogarth interview.

143 **Chinese Academy of Science projections:** Patrick E. Tyler, "China's Inevitable Dilemma: Coal Equals Growth," *New York Times,* October 25, 1995, p. D6.

143 **Global warming blown out of proportion:** Ross Gelbspan, "The Heat Is On," *Harper's*, December 1995, pp. 34–35.

144 **Lin Zongtang quote:** Patrick Tyler, "China's Inevitable Dilemma," p. A8.

145 **John Rogers quotes:** John Rogers interview.

145 **Safety was a big deal:** Ken Sghia-Hughes interview.

146 **Boston Edison spent $150,000:** James Hogarth interview.

146 **"Garlic with a twist of lemon":** Lynn Jacquez presentation at DARPA conference, May 1995.

147 **"My sex and age threw the uniformed guys":** Catherine Anderson interview.

148 **"Fuel cells sound almost too good":** David Swan interview.

148 **"The new guys have a lot to learn":** James Merritt interview.

149 **"We can't lobby!":** Bill Siegel interview.

149 **"The DOE spends huge sums":** James Worden interview.

149 **Shrinking technology funding:** A good example was the closing of the federal Office of Technology Assessment; science fiction writer Bruce Sterling in "The Future? You Don't Want to Know," *Wired,* December 1995, noted that government activism was "so entirely out of fashion that it's considered physically impossible. The easiest and cheapest way to restrict an overactive government is to blind it." He mocks this antitech attitude, saying, "So why worry about doubling the carbon dioxide in the atmosphere—in fact, why even keep records? If scientists at the National Oceanic and Atmospheric Administration proved beyond sane doubt that we were destroying the planet, we might feel obligated to try to do something. Why risk that possibility? Downsize the agency."

150 **Praising DOE role as key player:** John Gully interview.

151 **"Major Cope thought he could step in there":** James Worden interview.

152 **"Incongruity":** James Gleick, *Chaos: Making a New Science* (Viking Penguin, 1987), pp.

17. May 1995

158 **"Why aren't they out here":** James Worden in conversation.

158 **"I've only used one-third":** James Worden in conversation.

159 **"What that company needs":** Dave Dilts interview.

159 **"James wouldn't let me":** Don Walsh in conversation.

159 **Don Walsh quotes:** Don Walsh in conversation.

18. Crippling the Clean-Car Mandate

165 **"The anti-ZEV forces smell blood in the water":** Veronica Kun interview.

165 **"We're not ashamed of it at all":** Michael Parrish, "Trying to Pull the Plug," *Los Angeles Times*, April 14, 1994, Business section.

166 **Sheila Lynch quote:** Sheila Lynch interview.

166 **"The EPA doesn't have any guts":** Trudy Coxe interview.

167 **Governor Weld vacillates:** Scott Allen, "Weld Offers to Stall Rule on Electric Cars," *Boston Globe*, March 3, 1995, p. 17. Also Scott Allen, "Weld Key in Auto Emission Standoff," *Boston Globe,* July 29, 1995, pp. 1, 25.

167 **"Hey, Bill, if you're not running":** James Worden interview.

167 **Performance triggers:** Trudy Coxe interview, Anita Rajan interview.

167 **"The people in Trudy Coxe's office":** Anita Rajan interview.

167 **"Absolutely outrageous":** David Cohen interview.

168 **"The ZEV mandate in Massachusetts":** Trudy Coxe interview.

168 **Confidential memo from AAMA:** "Carmakers' Anti-EV Campaign in High Gear," CALSTART *Connection,* June/July 1995, p. 1, 3.

168 **"We're looking at this as a business":** Sean McNamara interview.

168–69 **PrEView facts and figures:** Bob Purcell, executive director of GM Advanced Technology Vehicles, in speech at North American EV and Infrastructure Conference, December 1995.

169 **"We are talking too much about EVs":** Noel Bureau, deputy director of research at Peugeot/Citroen, in speech at North American EV and Infrastructure Conference, December 1995.

169 **"Purely magical objects":** Roland Barthes, *The Eiffel Tower and Other Mythologies* (Hill and Wang, 1979).

171 **"Never before have so few been able":** Gregory Davis, *Technology Humanism or Nihilism*, p. 3.

171 **"Capitalism is myopic":** Lester C. Thurow, in review of *One World, Ready or Not: The Manic Logic of Global Capitalism*, *Atlantic Monthly*, March 1997, pp. 97–100.

171 **A strong, clear, consistent regulatory environment:** Jane Hall and others, "The Automobile, Air Pollution Regulation and the Economy of Southern California" (W. Alton Jones Foundation, 1995). Hall, a professor in the Department of Economics at California State University, Fullerton, writes in the study that the economy in the Los Angeles Basin grew much faster than that of the nation at large between the 1960s and the 1990s, "despite the area's more rigorous air pollution control environment." The implications? "Real progress occurs when political will exists over a sufficiently long period of time for the regulated community to seek effective new technologies. Consistent political support must be coupled with a solid scientific and technical foundation based on understanding the sources of pollution, the means to control them, atmospheric interactions, and the consequences of pollution."

172 **"I hear they're gloating":** Paul J. Miller in conversation.

172 **"The mandates are going down":** Sheila Lynch interview.

172 **"Don't get me going":** Ike Bayraktar in conversation.

172 **CARB folds on ZEV mandate:** Mark Rectin, "CARB Staff Urges End to 1998 ZEV Rule," *Automotive News,* December 25, 1997, p. 1.

19. Getting Sunrise into the Mainstream

175 **James Worden quotes:** James Worden interview.

176 **"A lot of people here really want":** Sheila Lynch interview.

176 **"A New England thing":** James Worden and Sheila Lynch interviews.

177 **Entries into EV industry:** The announcements appeared in press releases and elsewhere throughout the first months of 1996. Daihatsu, Hyundai, and other smaller automakers announced EV plans too. Samsung's announcement was detailed in CALSTART *News Notes,* June 6, 1997; the same notes describe the role played by the Korean Ministry ofTrade, Industry and Energy, which declared it would be helping Samsung, Hyundai, Dawwoo, and Hanjin Heavy Industries launch EV projects within the country.

178 **Populations on red alert:** The instances of cities experiencing bad air episodes were increasing. See CALSTART *News Notes,* May 30, 1997, about a study in New Delhi that said 7,500 people died prematurely in 1996 from pollution-related diseases; *News Notes,* August 31, 1996, reported that Hong Kong's air pollution had, on August 20, 1996, been the worst on record, with ozone levels far exceeding the safety level established by the city's environmental protection department.

178 **"Asia stinks":** Susan Moffat, "Asia Stinks," *Fortune,* December 9, 1996, p. 120.

178 **Timothy Wirth quote:** press release.

179 **Michael Oppenheimer quote:** Environmental Defense Fund interview.

179 **New Delhi, Shanghai phase out scooters**: "India Mulls Phase-Out of 2-Stroke Cycle Engines," CALSTART *News Notes*, May 30, 1997, p. 2; and "Shanghai Bans Mopeds," *ZAP News* (Sebastopol, California), Spring 1997, p. 2.

179–80 **Antarctic ice shelf disintegration:** Associated Press report, "Scientist: Antarctic Ice Shelf Near Collapse," *Burlington Free Press,* February 6, 1997, p. 2; in the report Rudi del Valle, director of geology of the Argentinian Antarctic Institute, said the 4,600-square-mile ice shelf was breaking up into thousands of icebergs, with cracks 30 miles long and 100 feet wide, and that two years before the northern section of the 620-mile-long Larsen Ice Shelf had collapsed.

180 **The U.S. continued to drag its feet:** John H. Cushman Jr., "U.S. Taking Cautious Approach in Talks on Global Warming," *New York Times,* December 8, 1996, sec. 1, p. 14; at the time, Cushman wrote, U.S. delegates at a United Nations meeting in Geneva argued that "it would be unrealistic to set new deadlines for reducing emissions of so-called greenhouse gases before the year 2010. And it is too early in the negotiations to specify how big any such reductions should be."

180 **Deb Callahan quote:** "Environmental Groups Say Gore Has Not Measured Up to the Job," *New York Times,* June 22, 1997, sec. 1, pp. 1, 16.

180 **Phillip Clapp quote:** *New York Times,* June 22, 1997, sec. 1, p. 16.

180 **Carolyn Browner quote:** H. Josef Hebert, "EPA Chief Defends Clean-Air Standards," *Burlington Free Press,* February 13, 1997, p. 6A.

181 **Jeff Fisher background:** Solectria Corporation résumés.

181 **Jeff Fisher quote:** Jeff Fisher interview.

182 **"At the end of the development program at Lotus":** Michael Jones interview.

182 **"If It really works, we'll do eight or ten more":** Wayne Kirk interview.

183 **"With composites, Solectria is getting where we know stuff":** Jeff Fisher interview.

183 **"Commercial voices are the ones":** James Worden interview.

183 **Worden on EV1 launch:** James Worden interview.

184 **"The market is global":** Jeff Fisher interview.

184 **Michael Jones quotes:** Michael Jones interview.

185 **"We don't know that it can't be done":** John Rogers interview.

Epilogue

187 **Jeff Fisher quote:** Jeff Fisher interview.

188 **"A niche within a niche":** James Worden interview.

188 **Force sales decline:** James Worden interview. (Worden said the decline had leveled off and the Force did continue to sell despite the competition.)

188 **Senator James Inhofe quote:** Associated Press, "Clinton Unveils Plan to Conquer Global Warming," *Maine Sunday Telegram*, February 1, 1998, p. 3A.

189 **"Do we need more evidence":** Ross Gelbspan, *The Heat Is On* (Addison-Wesley Publishing 1997), p. 150.

189 **"It would be unwise":** CALSTART *News Notes,* May 5, 1997, p. 2.

189 **Air pollution plague:** CALSTART *News Notes,* January 16, 1998, p. 2.

189 **EPA actions:** CALSTART *News Notes,* October 31, 1997, p. 2.

190 **Three quarters of new factories**: Keith Bradsher, "In the Biggest, Booming Cities, a Car Population Problem," *New York Times*, May 11, 1997, p. D4.

190 **Oil usage patterns:** James J. MacKenzie, "Driving the Road to Sustainable Transportation," *Frontiers of Sustainability* (Island Press, 1997), p. 133.

190 **Nineteen ninety-seven the warmest**: Randolph E. Schmid, "Climate Experts: 1997 Was Hottest Year Ever," *Burlington Free Press*, January 1, 1998, p. 3A.

191 **Jeff Fisher quote:** Jeff Fisher interview.

Index

Page numbers in *italics* refer to captions.

AC Propulsion, 195, 201
Aerocar, 16
AeroVironment, 50–51, 54–55, 56, 147, 170. *See also* General Motors, Impact/EV1; Sunraycer
  Joint Tactical Electrical Vehicle (JTEV), 147
Agile manufacturing, 135–136
Air toxics, 36, 44–45, 46
Alternative fuels, 60–62
American Automobile Manufacturers Association, 81, 95, 168
American Flywheel System AFS2000, 147, 148
American Lung Association, 84, 160, 198
American Solar Cup, 55
American Tour de Sol, 12, 15–17
  1989, 6, 56
  1990, 68
  1991, 70
  1994, 4–9, 116
  1995, 155–160, 162–163
  1996, 176–177
Anderson, Catherine, *18*, *30*, 31–33, 35, 49, 53–54
  after Solectria, 147
Antarctica, 179, 189, 211
Arizona Public Service Board, 71
  Electric 500 race of, 70
Art Center College of Design, 102
Arthur D. Little, Inc., 119
Asia, 178
Audi, 118, 177
Automobile Age, The (Flink), 140
Automobile industry, 140–142. *See also* Big 3 automakers; *specific manufacturers*
  and air pollution studies, 38, 42, 189
  and electric vehicle development, 23–25, 77–78, 129–131
  Hogarth's attitude toward, 111
  and ZEV mandate, 81–93, 166–169

Baker, Ken, 127–128
Barthes, Roland, 169
Batteries, 12, 20
  lead-acid, 64–65
  nickel-metal, 8, 155, 177
  sodium-sulfur, 7
  zinc-bromine, 72–75
Bayraktar, Ike, 172
Beaumont, Bob, 72, 124
Bell, Alexander Graham, 11
Big 3 automakers, 19, 149–150. *See also* Automobile industry; *specific manufacturers*
  objections of, to EVs, 21
  Worden's attitude toward, 106
Blair, Dave, 102, 104, 106, 117, 145
Blasch, Erik, *30*
BMW, 118
  E1, 14

Bodanis, David, 80
Boston, MA, 64
Boston Edison Company, 119, 146
and CitiVan, 188
and Sunrise Project manufacturing stage, 136, 137, 138, 150, 160
and Sunrise Project prototype stage, 7, 103, 106–107, 110–112, 158–159
Boxer, Barbara, 128
Brachos, Vasilio, 182
Bradley, Michael, 64, 80–81, 90
Braking, regenerative, 12, 100, 114
Brancazio, David, 31
Brancazio, Diane, *30*
Briscoe, Benjamin, 11
British Petroleum, 189
Browne, John, 189
Browner, Carol, 86, 180
Buchanan, Jamie, 89–90, 121
Bureau, Noel, 169
Bush, George, 63
Business Week, 20

Cabrillo, Juan Rodriguez, 39
California
early pollution concerns of, 46, 90, 196–197, 203
and ZEV mandate, *58*, 59–62, 78, 117, 160
California Air Resources Board (CARB), 41, 60, 65, 70, *164*
and ZEV mandate, 20–21, *22*, 172, 180
California Energy Commission, 62
California Institute of Technology, 37, 38, 41
California Motor Vehicle Pollution Control Act of 1960, 45
Caligiri, Joe, 32–33
Callahan, Deb, 180
CALSTART, 146, 187
Capitalism, Thurow on, 171
Carbon dioxide emissions, 36, 143, 178, 188
Carbon monoxide, 61
Carson, Rachel, 63
Catalytic converter, 47
Chaos: Making a New Science (Gleick), 152
Chapman, Ron, 86
Charles Draper Laboratory, 119, 138
Chevrolet S–10 pickup, 100, 163
China, 142–144, 179
Chinese Academy of Science, 143
Chrysler, 21, 24, 72, 103, 118, 120, 176
EPIC EV, 176
ESX, 176
TEVan, 124
Citroen. *See* PSA Peugeot-Citroen
Clapp, Philip, 180
Clean Air Act of 1963, 46, 62–63, 197
1970 amendment of, 45, 63, 91, 161, 189
1990 amendment of, 44, 63–64, 67, 71, 79, 90, 92, 161
Clinton, Bill, 84, 85, 120, 180, 188
Clinton administration, 178
Cohen, David, 167
Cole, David, 21
Composites, 20, 118, 137, 145–146, *174*, 182–183
Congress, 19–20, 91, 156–157, 161, 180
Conservation Law Foundation, 84, 87
Conversion cars, 4, 68–70, 100, 151, 152
Cope, Richard, 107–110, 112–113, 139, 151
Coxe, Trudy, 89, 121, 166, 168
Crockett, Don, 113

Daihatsu, 210
MP–4, 14
Daimler-Benz AG, 147
DARPA (Defense Advanced Research Projects Agency), 7, 77, 112, 149–150, 188
other programs of, 108–109
and Sunrise Project, 106–110, 120, 139, 146, 158, 183

Davenport, Thomas, 12
Davis, Gregory, 171
Delaware, 85
Dilts, Dave, 158, 159
Dockser, Mark, 137, 138, 181
Dow Chemical Company, 29, 55
Dow-United Technologies, 119, 137–138
Draper Laboratory, 119, 138
Dunlap, John III, 172
Dunlop, John B., 28
Durant, Billy, 140

"Early adopters", 11
Earth in the Balance (Gore), 86
Earth Summit, 1992, 188
"Ecolution", 14
Electric Fuel Corporation, 125
Electric vehicle industry, 3, 19, 23, 195
Electric vehicles (EVs) 12–13, 193–194. *See also* Hybrid vehicles; Solar cars
Electronic Power Technology, 108
Electrosource, 114
Ellis, Jim, 142
Emissions. *See* Clean Air Act of 1963; Smog; ZEV mandate
Energy security, 78
Environmental Defense Fund, 87, 179
Environmental Protection Agency (EPA), 45, 86, 119
  and California, 60
  creation of, 63
  and Ozone Transport Zone, 89–93, 121, 166, 180
EVermont, 108
EV Global Motors, 187
EVS-12 (World Electric Vehicle Association Symposium), 104, 117, *122*, 123–133

Fisher, Jeff, 181, 183, 184, 187, 191
Flink, James, 140
Flowers, Woody, 31
Flywheels, 20, 147, 148
Ford, 21, 24, 25, 72, 106, 158
  Ecostar, 5, 7, 9, 20, 116, 124, 163
  Model T, 13, 194
Ford, Henry, 140
France, 124
Freeman, David, 112
Fuel cells, 20, 147–148, 177

Gasoline, 15. *See also* Oil industry
  reformulated, 61, 80, 162
Gelbspan, Ross, 189
General Motors (GM), 21, 24, 38, 72, 106, 127–128. *See also* Sunraycer
  Chevrolet S–10 conversion, 100, 163
  Impact/EV1, 55, 64–65, 72, 124–127, 142, 168–169, 172–173, 176, 183
Geo Metro conversions, 4, 70, 100
Gleick, James, 152–153
Glidden, Charles Jasper, 11
Glidden Tour, 11–12, 15–17
Gliders, 100
Global warming, 143–144, 156, 178–180, 188–189
Gore, Albert, 86, 120, 180
Greenbaum, Daniel, 64, 84–85
Greenhouse effect. *See* Global warming
Gresens, Richard, *98*, 101
Ground-level ozone. *See* Smog
Grumet, Jason, 90
"Guerilla engineering", 32
Gully, John, 139, 150, 151

Haagen-Smit, A.J., 37–38, 40–42, 44
Hankinson, Scott, 71
Harris, Jay, 69, 70
Hayden, Tom, 85
Hazard, Nancy, 6
Heafitz, Andy, *30*, 155–156, 158
Health Effects Institute, 43, 198
Heat is On, The (Gelbspan), 189
Henault, Mark, *30*
Henderson, Douglas, 165
Higley, Lin, 4–5

Hogarth, James, 107, 110–111, 113, 119, 158
on China, 142
and Solectria management, 132, 135, 136, 159
Honda, 21, 69, 151, 177
Hong Kong, 210
Horlacher AG, 114
Fly, 113–114, 116
pickup, 148
H Power Corporation, 181
Hughes Aircraft Company, 50–51
Hybrid vehicles, 13–14, 96–97, 113, 116–118
Hypercar, 96–97
Hyundai, 210

Iacocca, Lee, 187
India, 83, 179, 210
Inhofe, James, 188
Institute for Economic and Environmental Studies, 171
Intergovernmental Panel on Climate Change, 143
Internal combustion engine (ICE) vehicles, 12–13, 14–17, 37, 176, 190. *See also* Automobile industry
Internet, 109
Inversion layer, 40, 197–198

Jacquez, Lynn, 146
John Paul Mitchell Systems, 50, 67
Johnson Controls, 73–74
Jones, Michael, 181–182, 184
Jorling, Thomas, 85

King, Angus, 161, 162
Kirk, Tisha, 71
Kirk, Wayne, 71, 102, 103, 106, 112, 115, 145, 182
and NIST grant, 137, 138
Kopec, Mark, 99, 169
Korea, 210
Kun, Veronica, 165
Kunin, Madeleine, 56
Kuo, James, 101–105, 106, 111, 115, 119, 125
after Solectria, 145
Kyoto conference, 188

Larson, Bruce, *30*
Latin, Howard, 91–93
"Least feasible retreat", 38, 42, 89, 149, 179
Lewis, C. Michael, 55
Lin Zongtang, 144
Liskowsky, Art, 6
Longley, James, 161–162
Los Angeles, CA, 46, 123
air quality in, 37, 38–42, 43, 178
Chamber of Commerce of, 38, 40
economy of, 171–172, 209–210
Los Angeles Herald, quoted, 39–40
Lovins, Amory, 96, 112, 117, 119, 127
Lowell, University of, 145, 182
Lynch, Sheila
and CitiVan, 188
and DARPA, 183
and NIST grant, 135–138
and Sunrise Project, 7, 107, 109–110, 111, 176
and ZEV mandate, 130, 172, 176

M113 personnel carrier, 113, 116
Maass, Dave, 137–138
McAlwee, Jerry, 114–115
MacCready, Paul B. Jr., 50, 54, 55, 62, 96, 129–130
on need for partners, 138, 140, 151
Machine That Changed the World, The (Womack), 135
McIntosh, David M., 161–162
McNamara, Sean, 126, 168
Maine, 160–162, 166
Mana La, 50, 51
Manchester Union, quoted, 15–16
Maryland, 85
Massachusetts, 21, 78, 81, 95, 160, 166–167
Massachusetts Electric Vehicle Demonstration Program, 109

Massachusetts Institute of Technology (MIT)
  Aztec, 7, 9
  Solar Car Racing Team of, 29, *30*, 31–33, 35, 49, 50
  Worden at, 3, 27–28, 115
Mazda, 21, 124
Meade, Gladys, 41
Mercedes-Benz, 147, 151, 177
Merritt, James, 148
Methanol, 60–62
Miller, Paul, 131, 172
Mirak Chevrolet, 28, 31
Mitchell, George, 161
Mitchell, Paul, 50
Mitsubishi, 118
  ESR, 14
Mobil Oil, 125, 127
Monga, Ritu, 156, 158
Morgan, Ted, 72
Muskie, Edmund, 161

National Institute of Science and Technology (NIST), 136–138, 150, 176, 182
New Hampshire, 85
New Jersey, 85
New York, 21, 78, 81, 166, 190
Nichols, Mary, 87
Nissan, 21, 124, 177
Nitrogen, oxides of (NOx), 36–37, 40, 198, 199
Northeast Alternative Vehicle Consortium (NAVC), 7, 107, 109–110, 150, 159
Northeast States for Coordinated Air Use Management (NESCAUM), 64, 80–81, 90
Northeast Sustainable Energy Association, 6. *See also* American Tour de Sol

Oil industry, 41–42, 125, 127, 165, 166
  and alternative fuels, 61–62
  and global warming, 143–144, 189
  and hybrids, 96
  and ZEV mandate, 81–83, 87, 166–169
Olds, R.E., 11
Oppenheimer, Michael, 179
Ovonics, 4, 8, 155, 177
Ozone burn, 38
Ozone Transport Commission, 79–88, 92, 95, 121
Ozone Transport Zone, *76*, 79
  and Maine, 160–162

Partnership for a New Generation of Vehicles (PNGV), 86, 97, 120, 148, 149–150, 156
Pentagon. *See* DARPA (Defense Advanced Research Projects Agency)
Persian Gulf War, 54, 77, 109
Peterson, Collin C., 161–162
Peugeot. *See* PSA Peugeot-Citroen
Pierce, Percy, 11
Pininfarina, 187
Pope, A.L., 11
Power, Virginia, 155
Pratt, Gill, *30*, 31, 68
Pratt, Janey, *30*, 73
PrEView Impact. *See* General Motors, Impact/EV1
PSA Peugeot-Citroen, 83, 124, 151
  106 EV, 169, 177
  AX, 169
Publicity, 16, 125–127, 146
Purcell, Bob, 169

Rajan, Anita, *30*, 53–56, 70, 115, 131
  on battery accident, 74
  and founding of Solectria, 3, 56–57, 67, 69, 71
  on hybrids, 97
  on performance triggers, 167
Rajan, Arvind, 73–74, 79, 95, 110, 115, 118, 137
Reagan, Ronald, 41
Reagan administration, 35, 47, 63

Reformulated gasoline, 61, 80, 162
Regenerative braking, 12, 100, 114
Regulations, government, 90–93, 170–172, 207–208. *See also* Clean Air Act of 1963; ZEV mandate
Renaissance Cars, 23, 72, 195, 201
  Tropica, 72, 124–125
Renault, 101
Rocky Mountain Institute, 96–97
Rogers, John, 145, 184–185
Rosen, Ben, 195
Rosen Motors, 195, 201

Sacramento Electric Transportation Consortium, 108
Sacramento Municipal Utility District, 71, 114
Safety, 28–29, 116, 145–146
Salmon, Norman, 65, 102, 104, 105, 106, 115
Samsung Electronics, 177, 210
Samuelson, Joan Benoit, 162
Sartre, Jean-Paul, 169
Saturn, 183
Schenk, Max, *174*
Schiffer, Michael Brian, 13
Schmitz, D., 50
Seal, Mike, 104
Sebring Vanguard Company, 72, 202
Sessa, Bill, 65
Sghia-Hughes, Ken, 69–70, 138, 145, 182
Sharpless, Jananne, 60–62, 65
Sherwin, Russell P., 43
Siegel, Bill, 149
Sierra Club, 84
Silent Spring (Carson), 63
Sloan, Alfred, 140
Smith, Dick, 50, 51
Smith, Jack, 120, 172
Smith, Roger, 64, 65
Smog, 36–37, 63–64
  early research on, 37–42
  effects of, 42–44, 82–83, 198
  in Northeast, 64–65, 161–162
Smuggler's Notch (VT) meeting, 112–116
Solar Car Challenge, 55
Solar cars, 5, *18*, 28, 116. *See also* Electric vehicles (EVs)
  building at MIT, *30*, 31–33
Solar Challenger (plane), 54–55
Solectria Corporation, 21, 23, 25, *52*, 78–79, 167, 168, 170, 191. *See also* Sunrise Project
  battery work of, 20
  founding of, 3, 56–57, 67, 69
  MacCready on, 130
  models of
    I and II, *10*, *26*, 27, 101
    III, 28–29
    IV, *48*, 49–51
    V, *1*, *2*, *18*, *26*, 53, 55, 56
    CitiVan, 187–188
    E-10, 7, 100, 108, 113, 188
    Force, 70–71, 73, *94*, 100, 109, 110, 163, 188
    Force RS, 4–9, 101
    Galaxy, 56
    Lightspeed, *66*, 67–68, 101
  shift to manufacturing of, 151–153
Southern California Edison, 71
Southern Coalition for Advanced Transportation, 108
Stanford Research Institute (SRI), 41, 44
State implementation plans (SIPs), 60
Stempel, Robert, 14–15, 112
Sterling, Bruce, quoted, 207–208
Sunrayce, 56–57
Sunraycer, 50–51, 54, 56
Sunrise Project, 96, *98*, 101–106, 116–119, *134*, 144–147, *154*, *174*, 175–176, 181–185, 187–188. *See also* Solectria Corporation
  and 1995 American Tour de Sol, 155–160, 162–163
  at EVS-12, 125–133

models of
P1, 150, 177
P3, 181, 182–183
P2001, 187
partners in, 7, 106–112, 119, 136–138, 175–176
Suntera Solar Chariot, 148
Supercar, 84–86, 120
Swan, David, 148
Swatch, 177
Swiss Tour de Sol, 28, 49–51
Systems approach, to projects, 115–116

Taking Charge: The Electric Automobile in America (Schiffer), 13
Taylor, Clark, 182
Team Marsupial, 50, 51
Technology forcing/following, 45, 131
Technology — Humanism or Nihilism (Davis), 171
Textron, Inc., 184
Tholstrup, Hans, 49
Thompson, William Irwin, 88
Thurow, Lester, 171
TNE II, 116
Tokyo Motor Show, 13–14
Tour de Sol. *See* American Tour de Sol; Swiss Tour de Sol
Toyota, 21, 124, 151
AP-X, 14
Prius HEV, 187
RAV4 EV, 176–177, 187
TPI Composites, 102, 184
Trembly, Ed, 27, 99, 105
and battery accident, 72–75
and conversion cars, 68, 69–70
Trotman, Alex, 120

Union for Concerned Scientists, 84
Unique Mobility, 23, 170, 187, 195, 201
United States Council for Automotive Research (USCAR), 72
United Technologies, 120
U.S. Electricar, 23, 72, 125, 139, 151, 195, 201

Vaaler, Erik, 31–33
Volatile organic compounds, 37, 40
Volkswagen, 118
Volvo, 83, 118, 177

Walsh, Don, 158, 159
W. Alton Jones Foundation, 131, 172
Warren, Jane, 43
Waxman, Henry A., 63
Weight, of vehicles, 118
Weinig, Sheldon, 128–129, 177
Weld, William, 167
Western States Petroleum Association, 165
Wills, Rob, 105
Wilson, Howard, 51
Wirth, Timothy, 178
Wogulis, Ed, 102, 104, 106, 145
Womack, James, 135, 136
Worden, Alexandra, *30*
Worden, James D., *2*, 53–56, 108, 167, *174*, 187
accidents of, 50–51, 72–75
in American Tours de Sol, 4–9, 155–156, 157–158, 160, 162–163, 177
on DARPA, 149
at EVS-12, 125–126, 131–133
first cars of, *10*, 27–29, 101
and founding of Solectria, 3, 56–57, 67, 69
on future of EVs, 19–20
at MIT, *30*, 31–33, 35, 49–51
on need for partners, 183
and NIST grant, 135–138
and Sunrise Project, 95–97, 103, 105–106, 139, 151–153
systems approach of, 115–116
Worden, John L. III, 29
Worden, John L. IV, *30*
Worden, Peter, 4, 5–6

World Electric Vehicle Association, Twelfth International Symposium of (EVS-12), 104, 117, *122*, 123–133
World Solar Challenge, 49–51, 56
ZEV (zero-emission vehicle) mandate, 65, 72, 78, 87–88, 119, 160, 180, 190
in California, 20–21, *22*, *58*, 62, 67, *164*
in Massachusetts, 95, 166–167